# Fundamentals of Microbiology Laboratory Manual

Eighth Edition

**Kendall Hunt**
publishing company

Marlene DeMers

Biology Department
San Diego State University

**Kendall Hunt**
publishing company

www.kendallhunt.com
Send all inquiries to:
4050 Westmark Drive
Dubuque, IA 52004-1840

Copyright © 2008, 2009, 2015 by Marlene DeMers.

ISBN 978-1-4652-2399-9

Printed in the United States of America

# TABLE OF CONTENTS

## PREFACE (COURSE DESCRIPTION)
All Editions

Fundamentals of Microbiology (Biology 211L) is a non-microbiology majors college course for pre-nursing and food and nutrition majors at San Diego State University. The goal is to teach basic microbiology techniques for allied health majors and to teach students how important microbes are in our daily lives. All of the laboratory methods are standard techniques taught in a major's microbiology class. The exercises have been chosen and adapted to the needs of non-microbiology major students. Safety is one of our primary concerns. Our purpose is provide students with sufficient information to understand why microorganisms behave as they do and the significance of their behavior in diagnostic, industrial and environmental microbiology. This manual is intended as a laboratory guide with specific procedures. It will be necessary for the student to use the lecture textbook and other references to accompany this manual. The format of the manual is designed to give learning objectives, the necessary materials, representative cultures, technical background, and procedures to guide the student through each exercise. There is a section to record results for each exercise. Questions that are meant to promote critical thinking have been included in the conclusions and discussions section. Questions will be added by the instructor as needed. Case histories have been included in the appendix, to also promote critical thinking. In general, the laboratory material is divided into three major sections. The first introduces general types of microbiology techniques. The second covers the medical portion of the course, using some of the general techniques from before, as well as new techniques. The last third covers water and food microbiology. The sections are meant to overlap. I hope that this manual will serve as a vehicle for students to obtain a better understanding of microbiology and how it affects our daily lives. Have fun!

## PREFACE (8TH EDITION)

The *Fundamentals of Microbiology Laboratory Manual* continues to be a work in progress. The Eighth Edition includes the newest safety instructions that the American Society of Microbiology has developed.

## ACKNOWLEDGMENTS

I want to thank Bruce Wingerd and the San Diego State University Biology Department for encouraging and giving me the opportunity to write this manual. I want to thank Janice Kaping for her help with some of the questions and Dr. Barbara Hemmingsen for her valuable contributions. I also want to thank all of the teaching assistants at SDSU who have taught Bio 210, for their valuable input for the second edition. A very special heartfelt thank you for my coworker, Tom Gibson. Without his computer knowledge, emotional support, and editing contributions, this manual would not have been written.

Special thanks to Dr. Barbara Hemmingsen for taking the time to edit the second edition and contribute to it with new information that was used in the third edition. A very special thank you to Mary Coleman for helping to type the third edition of the manual. I also want to thank Jon Rizzo from the San Diego State University Instructional Support Technical Department for his valuable computer support assistance.

Thank you to the Microbiology students who have used the first three editions and have made valuable contributions that have been included in the fourth edition. Also, thank you to Michael Leboffe, from San Diego City College, for giving me some valuable suggestions.

Acknowledgements for the fifth edition: Since this manual is a work in progress, a fifth edition was necessary. New photos have been added, along with information that could be used by non-major microbiology classes nationally. Special thanks to Dr. Jeanne Weidner for her valuable input.

Special thanks to Jon Fuller for believing in this lab manual and my dream of going national with it. The latest edition (6<sup>th</sup>) is still a work in progress, in the hope of accomplishing that dream.

8th Edition: Thank you to Gerardo Perez and Laurel Green at San Diego State University Biology Department.

Marlene DeMers

# SAFETY INSTRUCTIONS FOR MICROBIOLOGY LABORATORY CLASSES

## EQUIPMENT USED FOR DISPOSAL IN THE LABORATORY

**Laboratory Cart
Discard Pan(second shelf)**

**Step on Can with red
biohazard bag**

## ITEMS ON STUDENT BENCH

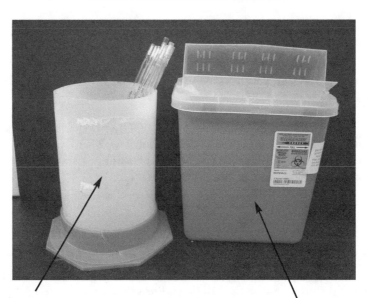

**Pipette Jar -
Tips down**

**Sharps Container
NO PAPER!**

**SAFETY NOTICE:** Once an item has been discarded into one of the containers described below, it must NOT be retrieved. The outside of the item is certain to have become contaminated, and the microorganisms will contaminate you! This rule will be strictly enforced.

## Disposal Equipment

### STEP-ON CAN
- Items placed in this can will be autoclaved (sterilized) and discarded.
- It is for **contaminated** paper towels, cotton, gloves, tongue depressors, and plastic Petri dishes. **Microscope slides do not go into this can.**
- All noncontaminated wood and paper products and cotton go in the regular trash.

### DISCARD PAN **(located on second shelf of laboratory cart)**
- Items in this pan will be autoclaved, washed, and used again.
- It is for reusable glassware such as test tubes, flasks, beakers, glass Petri plates, etc.
- **Remove all labels** from the intact glass but do not remove caps or liquid contents.
- Broken glass goes into the sharps container.

### SHARPS CONTAINER
The contents of the sharps container will be autoclaved and discarded. Discard slides, cover slips, broken glass, Pasteur pipettes, needles—that is, anything sharp whether contaminated or not. **NO PAPER!**

### PIPETTE JAR
Place reusable pipettes tip down into this jar. They will be autoclaved, washed, and resterilized.

## RULES FOR LABORATORY SAFETY AND SANITATION

**If these rules are not followed, you may eventually be asked to leave the lab permanently, and fail the lab.**

For the safety of everyone using the laboratories, it is REQUIRED that all members of each class use the equipment and follow the procedure as described below. There will be **quiz questions** on these safety and disposal rules. **Absolutely no eating, drinking, or smoking is allowed in the labs!! Lab coats must be worn in the lab at all times.**

## Safety Equipment

1. The **fire extinguisher** is near the hall door; use on objects only.
2. The **fire blanket** is in a labeled, red container next to the hall door; use the blanket to put out fires on people.
3. **Eye wash and shower** are by the hall door. Press the metal paddle, foot plate, or lanyard to start flow of water.
4. **First aid kits** are in each laboratory, by the stain shelves near the windows.
5. There is a **fume hood** available in NLS 420 for use with dangerous chemicals.
6. The **disinfectant** is stored in 5 gallon carboys in each lab near one of the sinks. Fill the disinfectant squirt bottle from this carboy and keep it near your work space.
7. **Liquid hand soap** is stored in dispensers over the sinks.

8. **Gloves** are located next to the glass cabinets or windows. Wear gloves when using chemicals (BSL) or at all times (BSL2).
9. **Material Data Safety Sheets (MSDS)** are located in a booklet on the stain shelves by the windows. Detailed copies are available at the Preproom window.

## Biosafety Levels

1. Most laboratory exercises will be performed with Risk Group 1 (RG1) organisms, using equipment and procedures designed for Biosafety Level 1 (BSL1). You will use the General Rules for Laboratory Safety on those days.
2. Certain laboratory exercises will be performed with Risk Group 2 (RG2) organisms (or environmental samples that could contain RG2 organisms), using special equipment and procedures designed for Biosafety Level 2 (BSL2). You will use the special BSL2 Rules for Laboratory Safety **and** the General Rules that do not conflict with the special BSL2 rules.

## Summary of Biosafety Levels for Infectious Agents

| BSL | Organisms | Procedures | Safety Equipment | Facilities |
|---|---|---|---|---|
| 1 | Not known to normally cause disease in healthy adults | Standard microbiological practices | PPE: Lab coats and eye protection | Disinfected lab bench and sink |
| 2 | Often associated with human disease | BSL1 practices plus: Limited lab access, Warning signs, Biosafety manual | PPE: Lab coats, eye protection, gloves Biosafety Cabinets for procedures that cause splashes or aerosols (flaming loops, pipetting) | Biosafety Cabinets, autoclave available |
| 3 | Disease may have serious or lethal consequences | BSL2 practices plus: Controlled lab access Decontamination of all discards leaving the lab Base serum collection | PPE: Lab coats, eye protection, gloves, respiratory protection Biosafety Cabinets for all procedures | Physical separation from hallways Self-closing doors Negative pressure room Room exhaust not recirculated |
| 4 | High risk of lethality, aerosol transmitted infections | BSL3 practices plus: Clothing change before entering, shower on exit | Class III Biosafety Cabinets or full-body, air-supplied, positive pressure suits | Separate building/zone Dedicated supply, exhaust, and decon |

## Safety Acknowledgement

1. Before you start any work in this class, you will be **required** to sign a Safety Acknowledgement sheet, which states:
   "I have read the Safety Instructions for Microbiology Laboratory Classes, San Diego State University, and I understand its content. I agree to abide by all laboratory rules set forth by this document and my instructor. I understand that my safety is entirely my own responsibility and that I may be putting myself and others in danger if I do not abide by all the rules set forth by this document and my instructor."

## Before You Take This Class

1. If you are **immunocompromised or immunosuppressed**, you may be at increased risk of acquiring infection. It is incumbent on you to discuss all such conditions with your personal physician prior to participation in the laboratory to determine if it is safe for you to participate in the laboratory exercises. Please bring a note from your doctor if he/she deems it safe.
2. If you are **pregnant**, you should also discuss this class with your personal physician prior to participation in the laboratory to determine if it is safe for you to participate in the laboratory exercises.
3. There is a **list of all organisms** used in this class at the end of these rules, for the information of your doctor and yourself.
4. You will be required to bring an **ink pen and a black Sharpie to lab, along with the specified lab notebook**. Bring a small zip lock bag, too. These will be disinfected or left in the lab at the end of the semester.

## General Rules for Laboratory Safety (BSL1)

1. **Disinfect your lab bench** before you begin work and just before you leave each day. Use the disinfectant squirt bottles and paper towels. Always assume that the lab benches are contaminated, even after disinfecting with disinfectant. The paper towels may be discarded in regular trash. Note: Although your squirt bottle may say "Amphyl," we now use a different disinfectant; the material safety data sheet (MSDS) is on the stock disinfectant carboy next to one of the sinks.
2. **Do not place books, backpacks, purses on the bench tops.** Always place these in the leg hole next to your seat. When you have started working on lab exercises, do not touch them again until you have washed your hands at the end of the laboratory.
3. **No cell phones** or other electronic devices (iPods, calculators, laptops, etc.) may be handled once the laboratory exercises have begun. They should be placed into your backpack or book bag and kept under the leg hole.
4. **Wash your hands** with soap before you begin work, before you take a break, and after you have disinfected your lab bench before leaving. Hand washing is the single most important thing you can do to reduce your risk of infection, so form this good habit early.
5. **Accidents and spills:** Report accidents and spills to the instructor immediately. If a spill has occurred, send someone to the instructor while you keep everyone away from the spill. If bacteria, blood, or fungi have spilled, cover the area with paper towels and gently flood with disinfectant. Wait 5 minutes, then clean up carefully. Do not handle broken glass with your hands. Place broken glass in the sharps container, and the paper towels in the step-on can. A dustpan and broom is located in each laboratory classroom and at the prep room window. Disinfect the dust pan afterwards.

6. **Keep your hair tied back** from your face and don't wear dangling jewelry.

7. **Eating and drinking** are prohibited in the lab at all times. Place any food or water bottles into your backpack or book bag **before** you enter the lab.
   **Keep your fingers and fomites (pencils, etc.) away from your eyes, nose, and mouth.** Do not touch your eyes or apply makeup in lab, including lip balm.

8. **No oral/mouth pipetting in the lab.** Use mechanical pipettors. Green pipettors are available at the instructor's bench. Be sure to return it when finished. Refer to the pipetting rules below.

9. **Do not remove media, equipment, or bacterial cultures** from the laboratory.

10. **Do not place contaminated instruments** such as inoculating loops and pipettes on bench tops. Loops and needles should be sterilized first, and pipettes placed into the pipette jar.

11. **Laboratory coats** covering the hips and the arms must be worn at all times in the lab. Keep this coat in your drawer. These coats must be sterilized before they are removed from the floor. The support staff in the prep room will do this for you.

12. **Use only the pens** you bought for this laboratory during the exercise. **Do not use any other writing instrument** during the lab exercises. If you wish to store these pens in your drawer, bring a plastic bag and place them in the bag first before placing them into the drawer. **Do not use these pens** during the preliminary lecture, use only your personal writing instruments. Remember to put these personal writing instruments away before beginning the exercises.

13. **Closed toed shoes** must be worn in the lab at all times. Sandals or flip-flops are not acceptable. If you want to wear sandals, etc., bring an old pair of shoes/sneakers and keep them in your drawer.

14. All procedures involving any human fluid (blood, urine, etc.), even your own, must be conducted behind **plastic shields**. Everyone must wear gloves (latex or vinyl) when working with blood or urine. If a shield is not available, you must wear safety glasses and masks.

15. All labs using hazardous chemicals (stains and reagents) should be performed using **gloves and eye protection**.

16. Non-latex gloves are available in each lab.

17. In the hematology laboratory you must also wear a lab coat impermeable to liquids.

18. Please put away your supplies and equipment at the end of a work session whether scheduled or non-scheduled. **Take great care with the microscopes**, they are delicate, precise, and expensive to repair or replace. Always return them to their correct cabinet position with slides removed, stages returned to the lowest position, and the lowest power objective in place. Clean your microscope (lenses and stage) with lens paper.

19. All cultures or materials must be properly identified with your name, the date, the identity of the culture or material, the exercise #, and the course number (or TA name). Items not completely identified will be autoclaved and discarded for the safety of all concerned. **Never discard anything that is not yours!** When in doubt, consult the instructor. The Sharpie label (no tape) should be written on the glass, not on the cap or white spot. Remove the label before discarding tubes into the discard bin.

20. Use the **waste disposal guidelines table** at the end of this document when there is a question where something should be discarded. This table is also in your lab manual and posted in the classroom laboratories.

21. Know what chemicals, stains, and reagents are dangerous and where to find MSDS information. This is in a booklet next to the stain shelves.

## Pipetting Rules

1. Note that pipetting can cause aerosols (droplet formation) that can spread microbes around the lab. To avoid this make sure to pipette **gently and slowly**, and **place the tip of the pipette against the container** before dispensing the liquid.
2. The glass 1, 5, and 10 ml pipettes should be removed from the metal container by shaking the container gently from side to side **until one pipette is extended** further than all the rest. Remove this pipette carefully, **without touching any other pipette** in the container.
3. Remember to **return the top** to the metal container when you are finished pipetting for this exercise (within 5 minutes or so).
4. Store the metal pipette canisters **on their sides**, not on end.
5. When pipetting using micropipettes, make sure to **close the top on the pipette tip box immediately** after you finish the exercise.
6. When removing microcentrifuge tubes from the beaker, **remove the foil lid carefully and aseptically, without tearing the foil.** Return the foil to the beaker without tearing it as well.
7. You can discard your micropipette tips and microfuge tubes into the beaker on your bench, but it **must be dumped** into the red bag before you leave.

## Special Rules for BSL2 Exercises

1. These rules are in effect whenever the printed lab schedule or your instructor specifies that today's exercise is a BSL2 exercise. All of the General Rules above also apply to these exercises, unless they are replaced by a BSL2 rule.
2. Place all **food, water, electronic devices, and other similar personal items** into your backpack or book bag **before you enter the lab.** When you enter, immediately place all personal items into the leg hole next to your seat. **Do not let them touch** anything else in the lab! Then **immediately decontaminate your bench top** with disinfectant, and wash your hands.
3. During the preliminary lecture period, you may bring out a pen and notebook or other items to write on after decontaminating the lab bench. Do not bring the backpack out of the leg hole! Before the exercise starts, all such items **must be returned** to your backpack.
4. **Your notebook or detached manual pages** and the **pens** you bought for this class when used during a BSL2 exercise may **not** be returned to your drawer or backpack (place them into your lab locker), and must be disinfected or kept in the lab at the end of the class.
5. The **doors and windows** must remain **closed and locked** at all times. **Only personnel trained in BSL2 procedures** may enter the laboratory when BSL2 exercises are being performed. Any other persons should remain outside the lab.
6. **Gloves** must be worn at **all times** during the exercises. Do not touch anything but the exercise materials, your locker items, and the bench top while wearing gloves. This includes any personal items (cell phone, keyboards, etc.). These gloves should be removed when finished with the exercises and placed into the red bag. Wash your hands after disposing of the gloves.
7. **Do not create any aerosols!** You may **not flame loops or do any pipetting** during a BSL2 exercise. You must also be careful not to create any splash or droplet formation, no matter how small.
8. To transfer cultures use only **sterile wooden sticks**, and **immediately** discard them into the waste beaker on your desk. You must use them carefully, and minimize their movement (bring the tube or plate to the stick, not the stick to the object). You **may not use your loops** or needles, and **no flaming** of any sort! Empty the waste beaker into the red bag at the end of the period.

9. If you are transferring from a liquid culture, **do not immerse the sterile stick more than ¼ of the way into the liquid.** If you immerse more of the stick, a droplet may fall from the stick during the transfer.

10. **Move your cultures to/from the incubator and refrigerator** only using a **test tube rack or beaker.**

11. At the end of the semester, you will **disinfect your wall locker and all of your lab locker contents** and the contents of the reagent shelf above your bench. Wet a paper towel with disinfectant and wipe down all objects. This will kill any potential BSL2 organisms that might have contaminated these objects.

12. Your ink pen and Sharpie can also be disinfected, or if you do not wish to keep them place into the bin provided. The notebook and any pencils or grease pencils will remain in the lab at the end of the semester (they are porous).

13. A biosafety manual is available in the laboratory.

14. A copy of *Biosafety in Microbiological and Biomedical Laboratories* (BMBL) is available in the laboratory.

## Special Lab Protocol for Biosafety Level 2 Lab Exercises

Note: This protocol is in effect whenever the printed lab schedule or your instructor specifies that the day's lab exercises are BSL2.

**At the beginning of the period:**
1. Before entering lab, place all food, water, electronic devices, etc., in a backpack and place in leg hole at your seat. DO NOT let any materials touch the bench top until it is disinfected.
2. Disinfectant the lab bench top immediately.
3. Wash hands.
4. Unlock lockers and drawers.
5. Take out lab manual and lab notebook (if used), pen or pencil, and place on clean lab bench for preliminary lecture period. The backpack stays in leg hole with all other materials left in it!

**After instructor announces that the BSL2 exercise will be starting:**
1. Take out a piece of notebook paper or tear out the results page from the lab manual (Bio 211L) or lab notebook (Bio350) (these will be used in the "dirty" area and kept in the lab locker).
2. Place both clean lab manuals in student drawer, with the one on top opened to the procedure for the day's lab exercise being done.
3. Place the clean pen or pencil in the "clean" drawer or backpack (for use later).
4. Place the sharpie that will become "dirty" with the notebook paper or results page, or lab notebook that is now on the lab bench.
5. Place all other materials in backpack under leg hole.
6. Put on gloves.
7. Begin the BSL2 exercise.

**Cleanup:**
1. Put away all cultures and lab materials.
2. Place "dirty" papers or lab notebook and Sharpie in lab locker.
3. Wipe lab bench with disinfectant; include anything that gloved hands touched, which includes the drawer handle.
4. Remove gloves.
5. Wash hands.
6. Remove clean lab manuals to backpack.

7.  Lock drawer and lab locker.
8.  Remove backpack and leave.

**To take home notes or results collected during the exercise:**
1.  Disinfect bench top.
2.  Put on one glove.
3.  Remove papers from lab locker, with gloved hand to put on bench top.
4.  Remove glove.
5.  Wash hand.
6.  Take picture of the paper with results with a cell phone or transcribe onto clean paper with clean pen or pencil (from clean drawer). Keep clean and dirty papers separate. (Note, Bio211L: it would be best to bring a photocopy of the results page from the lab manual.)
7.  Put on glove to move "dirty papers" back to lab locker.
8.  Disinfect lab bench.
9.  Remove glove.
10. Wash hands.
11. Lock drawer and locker.

## LIST OF ORGANISMS FOR BIO211L AND BIO350

| Microbe Name | BSL | ATCC # |
|---|---|---|
| Bacteria: | | |
| *Alcaligenes faecalis* | 1 | 8750 |
| *Azotobacter chroococcum* | 1 | 9043 |
| *Bacillus megaterium* | 1 | 14581 |
| *Bacillus subtilis* | 1 | 6051 |
| *Citrobacter farmeri* | 1 | 51113 |
| *Citrobacter freundii* | 1 | 8090, 43864 |
| *Citrobacter sedlakii* | 1 | 51115 |
| *Citrobacter werkmani* | 1 | 51114 |
| *Citrobacter youngae* | 1 | 29935 |
| *Clostridium sporogenes* | 1 | 3584 |
| *Corynebacterium pseudodiphtheriticum* | 1 | 10700 |
| *Corynebacterium xerosis* | 1 | 373 |
| *Enterobacter aerogenes* | 1 | 13048 |
| *Enterobacter cloacae* | 1 | 13047 |
| *Enterococcus faecalis* | 2 | 19433 |
| *Escherichia coli K12* | 1 | 2374 |
| *Escherichia coli K12* EM 499 (F'Tn10dTet) | 1 | - |
| *Escherichia coli K12* EM4 (Str$^R$) | 1 | - |
| *Geobacillus (Bacillus) stearothermophilus* | 1 | 12980 |
| *Hafnia alvei* | 1 | 13337 |
| *Lactobacillus bulgaricus* (yogurt) | 1 | 11842 |
| *Micrococcus luteus* | 1 | 4698 |
| *Moraxella catarrhalis* | 1 | 25238 |
| *Mycobacterium smegmatis* | 1 | 14468 |
| *Pantoea agglomerans* | 1 | 49174 |
| *Propionibacterium sp* (cyclohexanicum) | 1 | 700429 |
| *Propionibacterium acnes* | 1 | 6919 |
| *Providencia alcalifaciens* | 1 | 9886 |
| *Providencia rettgeri* | 1 | 29944 |
| *Providencia stuartii* | 1 | 29914 |
| *Pseudomonas aeruginosa* | 2 | 27853 |
| *Pseudomonas fluorescens* | 1 | 13525 |
| *Pseudomonas stutzeri* | 1 | 17588 |
| *Psychrobacter urativorans* (Micrococcus cryophilus) | 1 | 15174 |
| *Rahnella aquatilis* | 1 | 33071 |
| *Salmonella enterica serovar typhimurium* LT2 | 1 | - |
| *Salmonella enterica serovar typhimurium* AS 16 | 1 | LT2 |

| | | |
|---|---|---|
| *Salmonella enterica serovar typhimurium* trpD46 | 1 | LT2 |
| *Salmonella enterica MST 100* (Str$^R$) | 1 | LT2 |
| *Salmonella enterica MST 101* (Str$^R$ r-m+) | 1 | LT2 |
| *Salmonella typhimurium enterica* (LPS+)MST1757 | 1 | LT2 |
| *Salmonella typhimurium enterica* rfaG MST5023 | 1 | LT2 |
| *Salmonella typhimurium enterica* rfb MST2626 | 1 | LT2 |
| *Salmonella typhimurium enterica* rfc MST2627 | 1 | LT2 |
| *Serratia marcescens* (red) | 1 | 13880 |
| *Serratia marcescens* (white) | 1 | 6911 |
| *Staphylococcus aureus* | 2 | 12600 |
| *Staphylococcus epidermidis* | 1 | 14990 |
| *Streptococcus equi* | 2 | 33398 |
| *Streptococcus pyogenes* | 2 | 19615 |
| *Streptococcus salivarius* | 1 | 13419 |
| *Streptococcus thermophilus* (yogurt) | 1 | 19258 |
| *Streptococcus viridans* (group) | 2 | |
| *Streptomyces sp* | 1 | 19251 |
| *Streptomyces griseus* | 1 | 10137 |
| *Vibrio natriegens* | 1 | 14048 |
| **Fungi and Yeast:** | | |
| *Aspergillus* | 1 | |
| *Penicillium* | 1 | |
| *Rhizopus* | 1 | |
| *Saccharomyces cerevisiae* | 1 | |
| **Phage:** | | |
| **P22H int** | 1 | |
| **P22 H5** | 1 | |
| **Ffm** | 1 | |
| **Felix 01** | 1 | |
| **Protozoa, miscellaneous:** | | |
| Cyanobacteria with pond water (from greenhouse) | 1 | |
| *Euglena* | 1 | |
| Ground turkey (microbial spoilage of meat) | 2 | |
| Hay infusion | 1 | |
| Marine water samples from Mission Bay areas | 2 | |
| Pasteurized whole milk | 1 | |
| Plankton (marine) | 1 | |
| Raw whole milk | 2 | |
| Wine yeast (packaged, purchased) | 1 | |
| Yogurt w/ active cultures | 1 | |

## WASTE DISPOSAL GUIDELINES

| Disposal Item | Red Bag | Sharps Cont. | Pipette Jar | Glass Bin on Cart | Regular Trash |
|---|---|---|---|---|---|
| *Contaminated disposable plastic, paper, gloves, and gauze | X | | | | |
| Contaminated (used) plastic culture plates, taped closed | X | | | | |
| Contaminated (used) lens and bibulous paper | X | | | | |
| Microcentrifuge tubes | X** | | | | |
| Microliter pipette tips | X** | | | | |
| Urine and feces containers | X | | | | |
| Gloves used for routine protection or staining | X | | | | |
| Gauze, Kimwipes, etc., contaminated with blood | X | | | | |
| Contaminated applicator sticks and swabs | X** | | | | |
| Blood, blood vials or tubes | | X | | | |
| Slides, cover slips, and Pasteur pipettes | | X | | | |
| Needles, lancets and syringes | | X | | | |
| Other contaminated sharp items | | X | | | |
| Glass serological pipettes only (reusable) | | | X | | |
| Glass culture tubes, labels removed | | | | X | |
| Glass Petri dishes, labels removed | | | | X | |
| Uncontaminated paper: swab wrappers, Kimwipes, swabs, sticks, etc. | | | | | X |
| Paper towels used after hand washing and drying | | | | | X |
| Paper towels used for routine decontamination of bench before and after each lab period | | | | | X |

*Contaminated means any contact with all biological materials, such as: blood, urine, feces, throat, mucus, skin, bacteria, fungi, protozoa, and viruses.

**Use waste beakers, when full empty into red bag in laboratory classroom.

# General Microbiology

In the general microbiology section, you will learn the basic techniques and tools that are used in the microbiology laboratory. The microscope is the one of the most important tools used. It will be the first piece of equipment you will learn how to use, and you will continue to use it throughout the semester. It will be very helpful to master its use.

You will be introduced to a variety of microorganisms and learn the differences between protozoa, algae, cyanobacteria, fungi, yeast, and bacteria. This course does not include viruses, although they will be included in the lecture portion. After understanding and observing the basic differences between the procaryotes and eucaryotes, the course will mainly focus on the procaryotes, specifically, bacteria.

The exercises are arranged for the beginning microbiology student to learn the basic techniques for handling bacteria. This section will cover bacterial morphology and physiology. In some cases, only demonstrations will be used, such as in the miscellaneous staining exercise (capsule stain). Even though some of the procedures are introduced as demonstrations, the actual procedure has been included for laboratories wishing to have students perform the actual task.

During the staining exercises, you will learn the major characteristics of different genera of bacteria. The Genus, *Bacillus*, Gram-positive, endospore-forming bacteria is one example. Also, the Gram-negative bacteria, a major group of bacteria will be introduced in this section.

Safety: Most of the laboratory exercises in this section will be performed using Risk Group 1 (RG1) microorganisms, using equipment and procedures designed for BioSafety Level I (BSL1). The Medical section will have certain laboratory exercises that will be performed with Risk Group 2 (RG2) microorganisms using special equipment and procedures designed for Biosafety Level 2 (BSL2). The Environmental section will use environmental samples that could contain RG2 microorganisms and will be performed using BSL2 procedures. Each exercise will include the type of risk group with a list of the microorganisms and their biosafety level (BSL1 or BSL2) when necessary.

# EXERCISE 1

## The Brightfield Microscope

### OBJECTIVES

At the conclusion of the exercise, you should...

1. know the parts of the compound microscope and their purpose.
2. know how to safely transport the microscope.
3. be able to properly clean the microscope.
4. know how to store the microscope safely.
5. observe various specimens on microscope slides using the low, high, and oil immersion objectives.

### INTRODUCTION

The microscope is one of the most important tools that anyone studying microbiology must learn to use. During this lab exercise, you will learn the proper use of the brightfield microscope and you will practice using the microscope with some prepared slides and, later, with living specimens.

### MATERIALS

Materials for all exercises are listed per group (as determined by the instructor).

Safety: Biosafety Level 1

Supplies:

Prepared slides of various microorganisms
brightfield microscope. (See figure.)

Technical Background

**Parts of the Microscope:** (Refer to the figure of the Brightfield Microscope.)
1. Ocular eyepieces: magnifies the object 10-15X
2. Diopter ring: for adjusting the focus to the user's eyes
3. Rotating nosepiece: for rotating the objectives
4. Objectives: 5X, 10X (low), 40X (high dry), and 100X (oil) lenses that magnify the specimen
5. Stage: platform that holds the microscope slide and slide holder

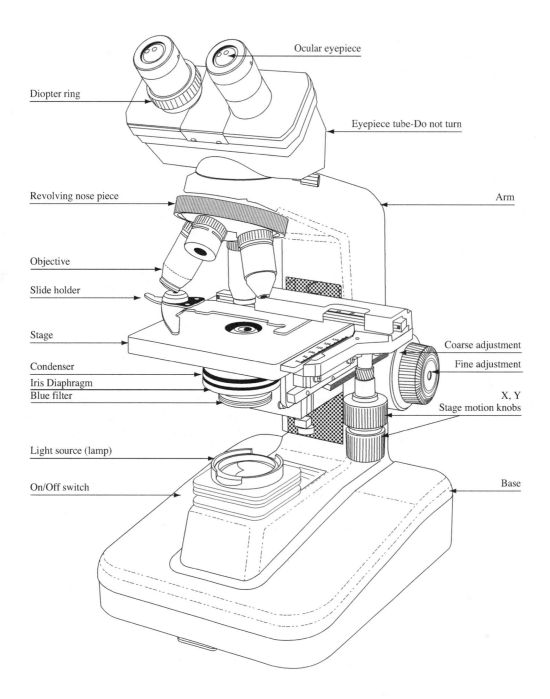

Diopter ring

Ocular eyepiece

Eyepiece tube-Do not turn

Revolving nose piece

Arm

Objective

Slide holder

Stage

Coarse adjustment

Fine adjustment

Condenser

Iris Diaphragm

Blue filter

X, Y
Stage motion knobs

Light source (lamp)

On/Off switch

Base

Brightfield Microscope.

6. Condenser: lens that condenses the light before it passes through the specimen
7. Condenser aperture diaphragm lever (iris diaphragm): controls the amount of light that passes through the condenser lens
8. Daylight filter: blue filter that provides a short wavelength for maximum resolution (omitted from some models of microscope)
9. Base
10. Light source: for power on and off, and controlling lamp brightness
11. X and Y stage travel controls: for moving the slide on the stage
12. Fine- and coarse-focusing knobs: for focusing the specimen
13. Arm: for carrying the microscope

## Microscopic Terminology

**Compound Microscope:** uses two (or more) lenses between the eye and the object.

**Brightfield Microscope:** the object being observed is dark against a bright field.

**Total Magnification:** is calculated by multiplying the magnifying power of the objective times that of the eyepiece. The ocular eyepiece is usually 10X and the objectives are 5X, 10X, 40X, and 100X. For example, when using the 40X objective and a 10X ocular, the total magnification would be: 40 x 10 = 400. Total magnification = objective magnification x eyepiece magnification.

**Resolving Power:** the ability of the lens to distinguish fine detail. Technically this refers to its ability to distinguish between two points a specified distance apart. The more minute the distance, the higher the resolving power of the optical system.

**Numerical Aperture:** used to determine the efficiency of objective lens to capture light. The larger the numerical aperture, the brighter and better resolved the image.

**Refractive Index:** a measure of the light-bending characteristic of a medium. As light passes through the air between the objective lens and the slide, it is bent because the refractive indices of these two media are different.

**Working Distance:** the distance between the specimen and the tip of the objective lens. In general, the higher the magnification, the shorter the working distance.

**Field of View:** the area or diameter of the specimen that is in view. The higher the power of magnification, the smaller the field of view.

**Depth of Field:** the thickness of the object that is simultaneously in focus. The higher the power of magnification, the smaller the depth of focus.

**Low Power Objective:** 5X or 10X. The objective used to locate the object on the slide, using the coarse adjustment knob.

**High-Dry Objective:** 40X. The objective used for higher magnification without oil.

**Oil Immersion Objective:** 100X. The objective used for the highest magnification with oil.

## PROCEDURES
### Day 1: Brightfield Microscopy Using Low and High-Dry Objectives

*Viewing a Specimen*

1. Transport the assigned microscope to the lab bench, using both hands.
2. Plug in the microscope and turn on the power. Make sure the rheostat is turned all the way up to allow maximum light.
3. Clean the ocular lenses and objectives with lens paper before use.
4. Place a prepared slide with the coverslip up on the stage within the spring-loaded lever.
5. Turn the rotating nosepiece until the 10X is above the ring of light coming through the slide.
6. Move the slide using the X and Y stage travel knobs until the specimen is within the field of view.
7. Bring the condenser up to the bottom of the slide and then slightly back for maximum light.
8. Adjust the ocular distance for your eyes by sliding the eyepiece plates in or out.
9. Adjust the focus by looking first into the right eyepiece and focusing the image with the coarse- and then fine-focus knob.
10. Look into the left eyepiece with the left eye, and focus the image by adjusting the diopter ring. The best focus is usually when the two white dots are aligned on the left ocular.
11. Rotate the revolving nosepiece to the high-dry objective.
12. Adjust the aperture diaphragm until there is sufficient light passing through the specimen. This will take practice. For a bright field, open the diaphragm completely. For a darker field, close the diaphragm slowly while observing the specimen. Generally, the higher the magnification, the more light will be needed to view the specimen.
13. Practice observing at least six prepared slides with the 10X and the 40X objectives.
14. When finished, remove the slide.
15. Clean the microscope oculars and objectives with lens paper. Remove any oil.
16. Rotate the 5X or 10X objective into place over the stage.
17. Lower the stage.
18. Turn off the power.
19. Unplug and wrap the power cord around the base under the stage.

## PROCEDURES
### Day 2: Brightfield Microscopy Using Oil Immersion Objective   X10

*Technical Background*

The most important objective used in microbiology is the oil immersion lens (100X). The oil is placed between the objective and the slide and is used to prevent the loss of light due to the bending of light rays (refraction) as they pass through air. This enhances the **resolving power** of the microscope.

1. Follow the procedure on Day 1, Steps 1-13.
2. After focusing with the high-dry objective, turn the 40X objective away enough to place a drop of oil on the slide. DO NOT LOWER THE STAGE.
3. Rotate the oil immersion objective (100X) into the oil, pass through, and return. This is to be sure there are no air bubbles between the objective and the oil.
4. Use only the fine focus to bring the object into focus.
5. Practice viewing the prepared slides at 10X, 40X, and then with oil.

6. DO NOT let oil get on any of the other objectives.
7. Always rotate the oil objective away before removing the slide.
8. When placing another slide on the stage, carefully start with the 10X or the 40X before going to oil.
9. Never use the coarse focus with the oil objective in place. The slide could break, and the objective could get damaged.
10. Observe at least 6 slides with and without oil.
11. Clean the oil off the oil objective with lens paper. Always clean all the objectives with clean lens paper.
12. Replace the microscope the proper way.

## EVALUATION OF RESULTS
## (EXERCISE 1: THE BRIGHTFIELD MICROSCOPE)

### Purpose
*Observation and/or Data*
Draw representative fields of the microscopic observations. Include the total magnification (Total Mag):

Specimen: _____     _____     _____

Total Mag.: _____     _____     _____

Specimen: _____     _____     _____

Total Mag.: _____     _____     _____

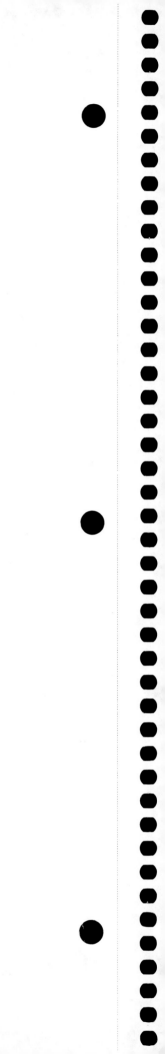

## CONCLUSIONS, DISCUSSIONS, AND QUESTIONS

1.  Why is oil necessary with the 100X objective?

    to enhance resolve strength

2.  If you are using an ocular that is 15X power and an objective of 40X, what is the magnification of the organism that you are viewing under the microscope?

    15×40= 600

3.  List the steps necessary for storing the microscope.

    1) clean and remove microscope slide
    2) rotate to 5x or 10x objective
    3) turn of andunblug, and wrap cord
    4) put away

# E X E R C I S E 2

## Other Microscopes

### OBJECTIVES

At the conclusion of the exercise, you should...

1. understand the difference between brightfield, darkfield, phase, and fluorescent microscopy.
2. know the purpose and characteristics of electron microscopes (EM).
3. understand the difference between a compound microscope and a stereoscope.
4. practice using the stereoscope.
5. observe demonstrations of wet mount specimens with the darkfield microscope.
6. observe demonstrations of wet mount specimens with the phase contrast microscope.

### INTRODUCTION

A variety of microscopes are available for different purposes in microbiology. The brightfield compound microscope is the most commonly used. In order to see a specimen clearly with a brightfield microscope, it has to have contrast. Contrast can be obtained by using stains. There are other types of microscopes that use more exotic means to generate contrast, such as darkfield and phase contrast. Unstained cells are more easily observed with these types of microscopes.

The stereo-type microscopes are commonly used as inspection instruments (stamps, coins, circuit boards) or as dissection microscopes. Their magnifying power is less than a standard compound microscope. The fluorescent microscope uses an ultraviolet source of light that causes the specially treated specimen to emit a fluorescent light. The fluorescent light observed is dependent on the type of fluorescent stain that is used.

In this exercise, you will observe examples of living microorganisms with darkfield and phase contrast microscopes. Later, the stereoscope will be used for observing the macroscopic structures of some fungi.

### MATERIALS

Safety: Biosafety Level 1

Cultures:

Various living microscopic organisms - BSL1
Petri dish with bacteria - BSL1

Supplies:

Prepared slides of various organisms
Darkfield microscope with wet mount of protozoa or algae
Phase contrast microscope with wet mount of protozoa or algae

## Technical Background

**Darkfield** microscopy is used to examine living microorganisms that are mostly invisible with a brightfield microscope. It uses a special condenser and reflected light. The reflected light causes the specimen to appear bright against a black background. It is most useful for viewing motile organisms that are difficult to stain.

**Phase Contrast** microscopy is also used to examine living microorganisms. It can help to distinguish the internal structures of some living specimens. It also uses a special condenser that allows both direct and reflected or diffracted light rays to come together to produce the contrasted image of the specimen.

**Fluorescent** microscopy is principally used to detect microbes in tissues or clinical specimens. It uses an ultraviolet light source that causes the microbes to give off a fluorescent light when stained with fluorescent stains.

**Stereoscope** (dissecting) microscopy is used in microbiology to observe larger specimens, such as fungi. It can also be helpful to observe the colony morphology of bacterial cultures. It uses two eyepieces (binocular) and two objective lenses to create a three-dimensional image. These instruments come in fixed, variable, and zoom designs, usually with a maximum power of 100X. The most common ones are between 20X–40X, with a large working area between the objective lens and the stage area.

**Electron Microscopy (EM)** uses a beam of electrons instead of light. It is used to examine objects smaller than 0.2 µm (micrometer). **TEM** (*transmission electron microscope*) uses electrons to pass through a thin section of material. **SEM** (*scanning electron microscope*) provides three-dimensional views of specimens without sectioning them.

## PROCEDURES

### Working with Microscopes

1. Use the darkfield microscope to observe the wet mounts prepared by the instructor.
2. Use the phase contrast microscope to observe the wet mounts prepared by the instructor.
3. Practice using the stereoscope at the lab bench by placing different objects on it.
4. Use the stereoscope to observe the plates with bacterial colonies.

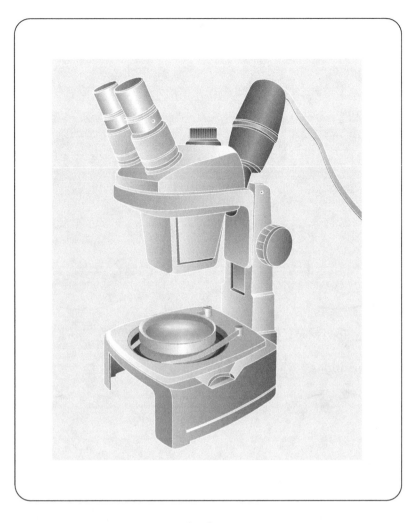

Example of Stereoscope.

| Table 2.1 | Summary of the Different Types of Microscopy and Their Applications in Microbiology | | |
|---|---|---|---|
| **TYPE OF MICROSCOPY** | **MAGNIFICATION** | **SPECIMEN APPEARANCE** | **APPLICATIONS** |
| BRIGHT FIELD | 1,000 TO 2,000 | STAINED OR UNSTAINED; BACTERIA ARE GENERALLY STAINED TO ENHANCE CONTRAST AGAINST BACKGROUND. | FOR GROSS MORPHOLOGICAL EXAMINATION OF BACTERIA, YEASTS, MOLDS, ALGAE, AND PROTOZOA |
| DARK FIELD | 1,000 TO 2,000 | GENERALLY UNSTAINED; APPEARS BRIGHT OR "LIGHTED" AGAINST A DARK BACKGROUND. | FOR SITUATIONS WHERE HIGH CONTRAST, PARTICULARLY OF MICROBIAL STRUCTURES, IS REQUIRED |
| FLUORESCENCE | 1,000 TO 2,000 | FLUORESCENT | FOR DIAGNOSTIC TECHNIQUES WHERE THE FLUORESCENT DYE (ANTIBODY) FIXED TO THE ORGANISM IDENTIFIES IT |
| PHASE CONTRAST | 1,000 TO 2,000 | CONTRASTING DEGREES OF DARKNESS (POSITIVE CONTRAST) OR BRIGHTNESS (NEGATIVE CONTRAST). | FOR EXAMINATION OF MINUTE CELLULAR STRUCTURES AND DETAILS IN LIVING CELLS |
| ELECTRON | 100,000 AND HIGHER | VIEWED ON A SCREEN | FOR EXAMINATION OF ATOMIC AND CELL STRUCTURES AND VIRUSES NOT OBSERVABLE WITH THE LIGHT MICROSCOPE |

# EVALUATION OF RESULTS
# (EXERCISE 2: OTHER MICROSCOPES)

## Purpose
*Observation and/or Data*

**Darkfield observations:**

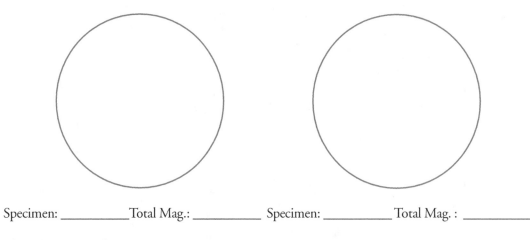

Specimen: _____ Total Mag.: _____        Specimen: _____ Total Mag. : _____

**Phase Contrast observations:**

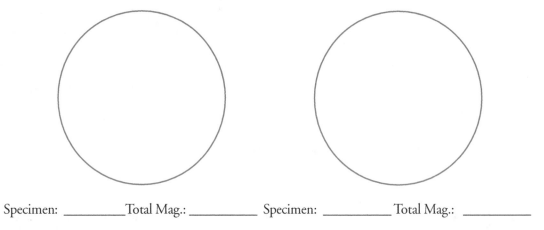

Specimen: _____ Total Mag.: _____        Specimen: _____ Total Mag.: _____

**Stereoscope observations:**

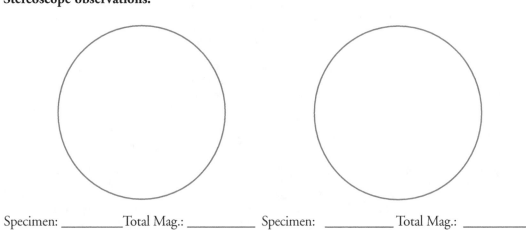

Specimen: _____ Total Mag.: _____        Specimen: _____ Total Mag.: _____

## CONCLUSIONS, DISCUSSIONS, AND QUESTIONS

1. What is the principle of, and principal use for, the darkfield microscope?

   - Uses reflected light and special condenser.
   - can view unstained.

2. What is the principle of, and principal use for, the phase contrast microscope?

   - Uses special condenser that uses the light rays combined to produce contrasted image
   - for cellular structures, details in living cells, internal structure

3. Explain how the appearance of the various living microorganisms differed with the dark-field, phase contrast, and brightfield microscopes.

   - Darkfield is for lighter, unstained organisms that are invisible to Brightfield. Brightfield is for stained and higher contrast.
   - Phase contrast is for much smaller structures.

4. Explain how the stereoscope microscope is useful in microbiology.

   Stereoscopes help observe larger specimen like the morphology of bacterial colonies in 3D.

# E X E R C I S E 3

## Observing Protozoa, Algae, and Cyanobacteria

### OBJECTIVES

At the conclusion of the exercise, you should...

1. be able to prepare wet mounts of various living microorganisms.
2. observe microscopic features of protozoa, algae, cyanobacteria, and bacteria.
3. recognize the diversity of microorganisms.
4. understand the basic differences between Procaryotes and Eucaryotes.
5. understand what hay infusion is.
6. be introduced to common genera and species of Eucaryotes.
7. know what a parasite is.

### INTRODUCTION

The microbial world includes a wide variety of microorganisms. Samples taken from ponds and oceans will have many different species of protozoa, algae, plankton, diatoms, and cyanobacteria. This exercise will give you more practice using the microscope and introduce you to the diversity of the microbial world.

### MATERIALS

Safety: Biosafety Level 1

Cultures: BSL1

Hay infusion
*Euglena*
Plankton tow
Cyanobacteria
Pond water
Plates of bacteria: BSL1
Taped: BSL2

Supplies:

Slides and coverslips
Lens paper

Pasteur pipettes with bulbs
Prepared slides of some medically important protozoa

# PROCEDURES

## Technical Background

**Hay Infusion:** An infusion is made by soaking dried plant material in water. For example, after soaking hay in non-chlorinated water for a week, a "hay infusion" will develop, yielding a variety of microorganisms. Protozoa are typically found in water, but when they become dry, they can form cysts and go into a dormant state. These cysts can be activated by adding water. At the beginning of the infusion, bacteria will predominate. Then a variety of other microorganisms will appear; these will include saprophytic flagellates, ciliates, and amoebae that feed on the bacteria. Eventually, the carnivorous ones will appear. The appearance of each new species is related to factors such as light intensity, gases present, pH, and concentration of organic compounds. The sequence of appearance of organisms in the hay infusion is representative of a food chain.

**Succession:** Succession refers to orderly sequential changes in the composition or structure of an ecological community. Succession in a protozoan community may be demonstrated in pond water.

**Procaryote:** A cell whose genetic material is not enclosed in a nuclear envelope and usually has a single circular DNA molecule as its chromosome. The bacteria are in this group.

**Eucaryote:** A cell having DNA inside a distinct membrane-enclosed nucleus (true nucleus) and usually other organelles. The protozoa and fungi (as well as animals and plants) are in this group.

**Protozoa:** Mostly unicellular eucaryotic microorganisms that lack cell walls.

**Parasite:** A type of organism that feeds on live organic matter, such as another organism.

**Algae:** The common name given to a heterogenous group of plants that are capable of carrying on photosynthesis and usually live in water. Some are unicellular.

*Euglena:* A one-celled organism that is usually green in color, can generally make its own food by photosynthesis (phototrophic), and is free moving.

*Tetrahymena:* A protozoan possessing cilia for motility.

**Cyanobacteria:** Oxygen-producing procaryotes; also called blue-green algae.

**Plankton:** Free-floating aquatic microorganisms. This group includes diatoms and dinoflagellates. Diatoms have cell walls that consist of pectin and silica.

*Making Wet Mount Slides*

1. After viewing the videos showing some of the microorganisms that can be found in the environment, examine the various samples to try to find examples.
2. Place a drop of one of the liquid specimens on a microscope slide, cover with a coverslip, and examine with the brightfield microscope, starting with 10X and going up to 40X.
3. Use the manuals and diagrams provided, as a guide.
4. Note the means of motility (flagella, cilia, ameboid).

5.  Note the difference in sizes of the bacteria and the protozoa. Protozoa can range in size from 1 mm to 70 mm or larger. Bacteria usually range from 0.5-2.0 µm.

6.  Practice making wet mounts of all the samples, and record the observations in the Evaluation of Results.

**Examples of Protozoa, Algae and Cyanobacteria**

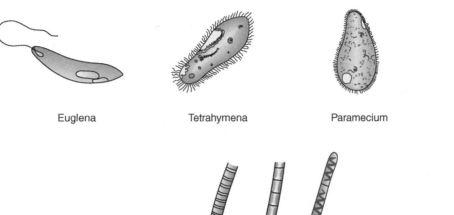

Euglena            Tetrahymena            Paramecium            Amoeba

Algae

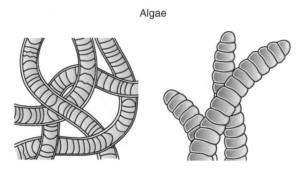

Cyanobacteria

Examples of protozoan parasites

 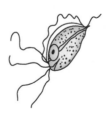

Giardia trophozoite            Trichomonas vaginalis

Examples of Protozoa, Algae, and Cyanobacteria.

## EVALUATION OF RESULTS
## (EXERCISE 3: OBSERVING PROTOZOA, ALGAE, AND CYANOBACTERIA)

Purpose
*Data*

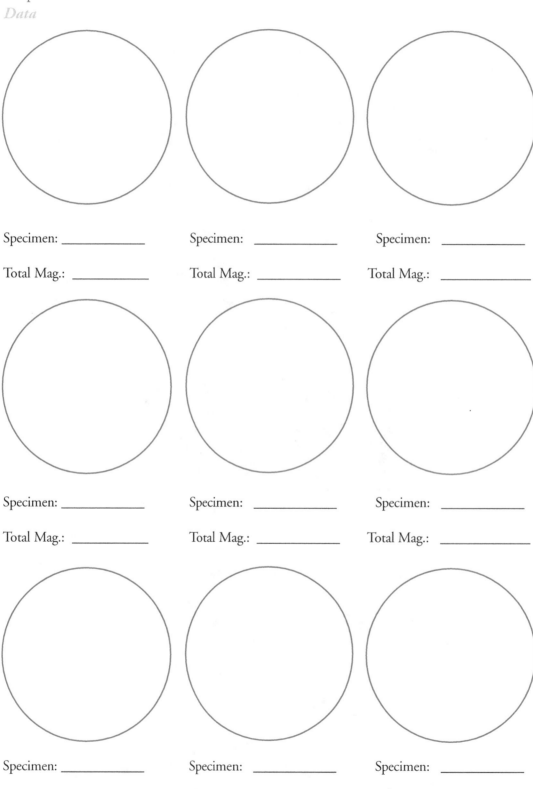

Specimen: _____          Specimen: _____          Specimen: _____

Total Mag.: _____         Total Mag.: _____         Total Mag.: _____

Specimen: _____          Specimen: _____          Specimen: _____

Total Mag.: _____         Total Mag.: _____         Total Mag.: _____

Specimen: _____          Specimen: _____          Specimen: _____

Total Mag.: _____         Total Mag.: _____         Total Mag.: _____

## CONCLUSIONS, DISCUSSIONS, AND QUESTIONS

1.  Explain two differences between protozoa and algae and two differences between algae and cyanobacteria.

    Protozoa

2.  Use your textbooks and the Internet to complete the table below.

    Table 3.1 Examples of Human Pathogens that are Protozoa.

| Name of pathogen | Disease | Source of human infections |
|---|---|---|
| *Giardia lamblia* | | |
| *Entamoeba histolytica* | | |
| *Naegleria fowleri* | | |
| *Cryptosporidium* species | | |
| *Trichomonas vaginalis* | | |

3.  Write a short paragraph in which you compare what you can see in the pond water or plankton tow with your naked eyes with what you can see through the microscope at 100X and 400X total magnification.

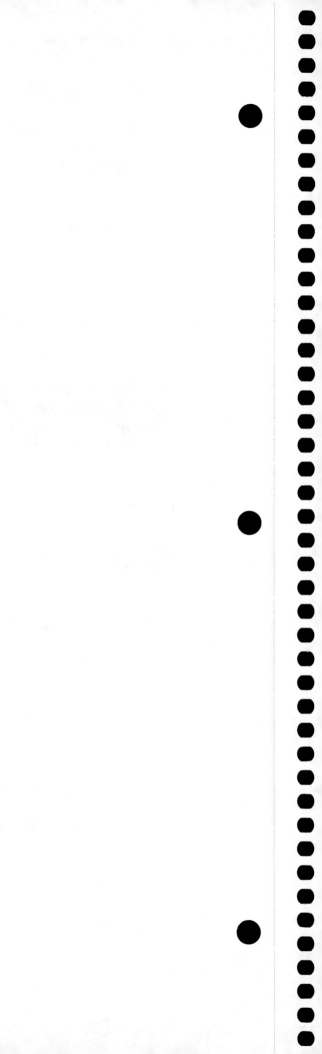

# EXERCISE 4

## Observing Fungi and Yeast

### OBJECTIVES

At the conclusion of the exercise, you should...

1. understand where the fungi are placed in the classification of microorganisms.
2. be able to recognize the characteristics of fungi to identify certain species of fungi macroscopically and microscopically.
3. observe the differences between bacteria and fungi growing on agar plates.
4. know the types of media that are used for growing fungi.
5. observe wet mounts of yeast.
6. be introduced to the concept of **compromised host**.
7. know the difference between a **pathogen** and an **opportunistic pathogen**.

### INTRODUCTION

Fungi (molds) are a diverse group of single-celled or multicellular organisms that obtain food by direct absorption of nutrients. Fungi are classified with the eucaryotic microorganisms. They are identified by their reproductive spores from the asexual and sexual stages of their life cycles. They play a very beneficial role in the food chain by decomposing dead plant matter, providing food (mushrooms), and aiding in the production of foods (bread) and drugs (alcohol). Also, over the last 10 years, there has been an increase in fungal infections in people with compromised immune systems. Thus, identification of a fungus is often necessary for the treatment of a disease. The study of fungi is called Mycology. In this laboratory, you will observe asexual and sexual structures of different species of fungi and learn how these characteristics aid in identification.

### MATERIALS

Safety: Biosafety Level 1

Cultures: prepared slides: BSL1

*Penicillium* species (phialospores/conidiospores)
*Aspergillus* species (phialospores/conidiospores)
*Rhizopus* species (sporangiospores)
*Rhizopus* species (zygospores)
*Saccharomyces cerevisiae* (yeast)
*Candida albicans* (yeast) - BSL2 - prepared slides only
Plates of bacteria for comparison - BSL2

Supplies:

Prepared microscope slides of the fungi and yeast listed

## PROCEDURES

### Technical Background

The **thallus** is the body of a fungus, and it consists of long filaments of cells joined together, called **hyphae** (singular: *hypha*). Hyphae can contain cross walls called **septa** (singular: *septum*) and are called **septate** hyphae. In a few classes of fungi, the hyphae will not contain septa, and are called **non-septate hyphae** or **coenocytic hyphae**. These filamentous fungi can reproduce both sexually and asexually by the formation of **spores**. Two kinds of asexual spores are produced: conidia or **conidiospores**–formed on specialized hyphae called **conidiophores**–and sporangia or **sporangiospores**, which are formed within a **sporangium (sac)** at the end of an aerial hypha called a **sporangiophore**. There are several types of conidiospores. **Phialides** are vase-shaped cells that produce chains of conidia called **phialospores**.

**Sexual** spores result from sexual reproduction where (+) and (-) nuclei fuse to form sexual spores. Three types of sexual spores are produced in fungi, **zygospores** (inside a zygote), **ascospores** (inside an **ascus**), and **basidiospores** (inside a **basidium**).

Fungi are classified into four categories: **Zygomycota**, **Ascomycota**, **Basidiomycota**, and **Deuteromycota**.

- **Zygomycota** have nonseptate hyphae and produce asexual sporangiospores. An example is *Rhizopus*, a common bread mold.

- **Ascomycota** produce conidiospores. Examples are *Aspergillus* and *Penicillium*.

- **Basidiomycota** have septate hyphae and produce conidiospores. Mushrooms are in this division.

- **Deuteromycota** is a "holding place" for fungi if they have not yet been found to produce sexual spores. It is sometimes referred to as the "imperfect fungi." It is a large group, containing over 15,000 species. Many of the pathogenic fungi are in this division.

**Yeasts** are classified as unicellular fungi that are non-filamentous. Yeast reproduce by either **fission** (divide evenly to produce two new cells) or **budding** (parent cells form a bud that eventually breaks off). When buds are produced and fail to detach, they form a chain of cells called a **pseudohypha**. Two examples of yeast are *Saccharomyces cerevisiae* (bread-making yeast) and *Candida albicans* (opportunistic pathogen).

*Candida albicans* is a **dimorphic** yeast, meaning that it takes on two forms. Most of the time it exists as single, oval-shaped yeast cells, which reproduce by budding. Under the right conditions, *Candida albicans* can develop **pseudohyphae**, which are composed of a chain of cells and grow as irregular filaments. Room temperature (25°C) yields the single and budding cells. Body temperature (37°C) and body pH, along with the presence of serum, allows *C. albicans* to be capable of producing pseudohyphae. A rapid identification test to identify whether a yeast colony is *C. albi-*

*cans* is a "Germ Tube" test. Human serum is inoculated and incubated at 37°C for 2-3 hours. The presence of pseudohyphae along with no pseudohyphae at room temperature gives a presumptive identification of *C. albicans*. Further biochemical tests can be performed. *C. albicans* can be found in normal humans in the mouth, gut, and vagina. Infections occur when a patient has become immunocompromised. Infections caused by *C. albicans* are called candidiasis or "thrush." Candida can also cause pneumonia, septicemia, and endocarditis.

*Saccharomyces cerevisiae* are single-celled fungi that multiply by budding or by division (fission). *S. cerevisiae* is known as the "bread-making" yeast because growth in the dough produces $CO_2$, causing the dough to rise. It is also used for wine and beer production. It is a member of the Ascomycota (ascus-forming) fungi. *Penicillium* and *Aspergillus* also produce these types of spores.

**Culture Media**: There many different types of culture media used for growing fungi. Potato Dextrose Agar (**PDA**), or broth, and Sabouraud's Agar (**SAB**), or broth, are two of the most commonly used. SAB culture media contains an antibiotic to inhibit bacterial growth. PDA consists of a potato infusion and dextrose, and will encourage the growth of fungi and yeast.

**Compromised host**: a person whose immune system is suppressed and, therefore, can be susceptible to invasion by microorganisms. A suppressed immune system can result from various diseases, therapies, or burns.

**Opportunistic pathogen**: microorganisms that invade the body when the body's immune defense is suppressed or compromised.

**Pathogen**: any microorganism that can cause disease in a healthy person.

*Viewing Fungi and Yeast Samples*

1. Examine the Petri plates of the fungi provided (please, do not open the plates).
2. Describe the color of the thallus and hyphae.
3. Describe and draw the colony characteristics for the front and back of the colony.
4. Compare the cultures to the drawings and photos provided.
5. Examine the prepared slides of the fungi provided.
6. Describe the type of hyphae and spores seen.
7. Make wet mounts of the yeast provided, and note the budding or presence of pseudohyphae.
8. Use the Evaluation of Results section to make drawings and write descriptions.
9. Observe plates of yeast and compare to plates of bacteria.

## Examples of Fungi

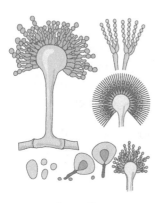

Aspergillus

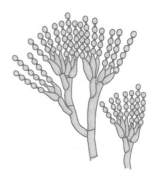

Penicillium

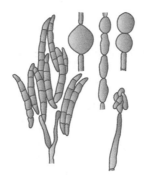

Fusarium

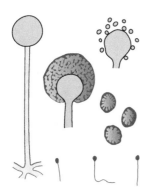

Rhizopus sporangia
(asexual spores)

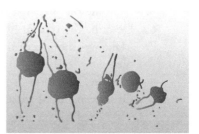

Rhizopus zygospores
(Sexual spores)

Examples of Fungi.

## Examples of Yeast

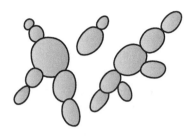

*Saccharommyces cerevisiae*

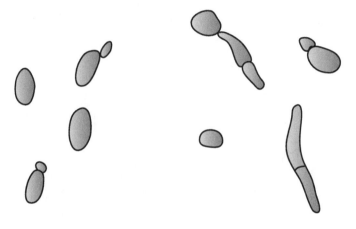

*Candida albicans*
Single and budding cells and pseudohyphae

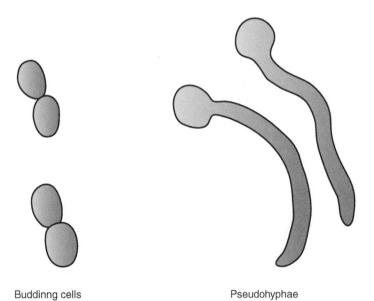

Buddinng cells                    Pseudohyphae

Examples of Yeast.

## EVALUATION OF RESULTS
## (EXERCISE 4: OBSERVING FUNGI AND YEAST)

Purpose
*Data*
Describe in your own words the fungal colonies growing on agar medium. Label the drawings to include color, type of growth (wrinkled, smooth, etc.). Then examine the prepared slides of each fungus. Draw the arrangement of hyphae and spores for each fungus.

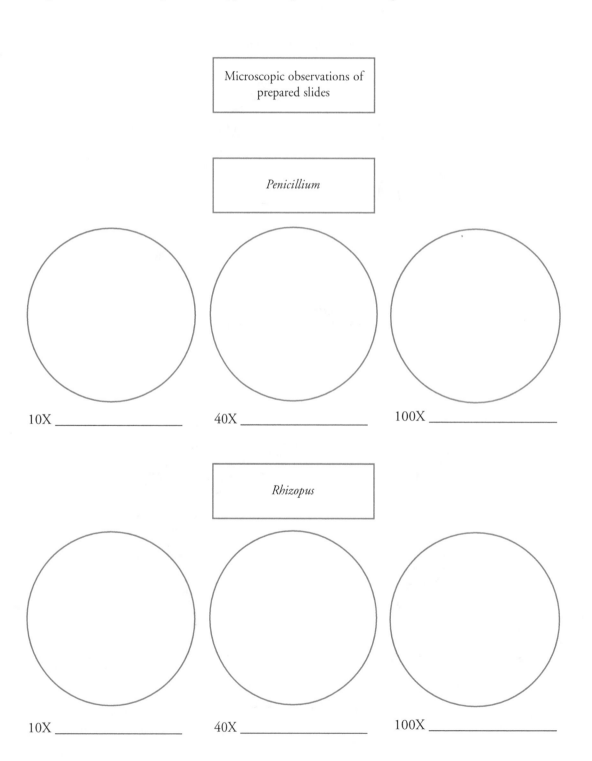

Microscopic observations of
prepared slides

*Aspergillus*

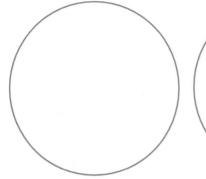 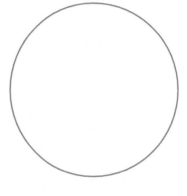

10X _____       40X _____       100X _____

*Candida*

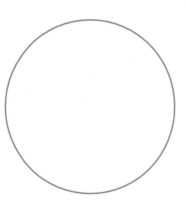  

10X _____       40X _____       100X _____

Wet mount of yeast *Saccharomyces cerevisae cerevisae* (400X)

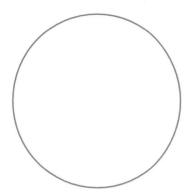

Stained slide of yeast *Saccharomyces*

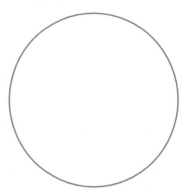

## CONCLUSIONS, DISCUSSIONS, AND QUESTIONS

1. Which slide preparation of yeast, the wet mount or stained slide, is easier to view? Why?

2. Match the following:

Eucaryote                                          a. science that studies fungi ✓

Mycelium                                           b. type of cell fungi have

Pseudohyphae                                       c. chain of yeast blastospores

Mycology a                                         d. mass of mold filaments visible
                                                      without a microscope

Dimorphic                                          e. type of hyphae seen in *Penicillium* ✓

Macroconidia                                       f. fungi with yeast and mold stages

Septate hyphae e                                   g. type of hyphae seen in *Rhizopus* ✓

Non-septate hyphae g                               h. asexual spores of *Fusarium*

# EXERCISE 5

## Observing Bacteria

## OBJECTIVES

At the conclusion of the exercise, you should

1. recognize the differences between bacteria and other organisms studied so far.
2. know the different types of microscopic morphology of bacteria.
3. be able to recognize the shapes of bacteria.
4. be able to recognize the sizes of bacteria.
5. become familiar with observing bacteria that have been stained.

## INTRODUCTION

The shapes of bacteria can be grouped into three types: bacillus (rectangular or rod shaped), coccus (spherical or round), and spirillum (curved or helical). There are variations within each of these shapes. In this exercise, you will continue practicing using the microscope to examine the variety of shapes and sizes of bacteria.

## MATERIALS

Safety: Biosafety Level 1

Cultures: BSL1

Prepared microscope slides of a gallery of bacterial shapes and sizes

## PROCEDURES

### Technical Background

Most bacteria range in size between 0.5 and 2.0 micrometers (μm). One aspect of bacterial identification is to describe their shape. The three basic types are rod, spherical, and curved. Rods can vary in length and width; may have square, round, or pointed ends; may or may not have flagella (for motility); and may occur singly or in chains. Rods that have variable sizes among individual cells (i.e., coccoid to long rods) are referred to as pleomorphic in shape. The spherical coccal cells may occur singly, in pairs, tetrads (groups of four), chains, or irregular clusters. The helical or curved bacteria can vary in length and width, and may occur as curved, bent, or wavy forms, with or without flagella. There are bacteria that form endospores, and this morphological characteristic is also used in classifying bacteria.

The following are examples of the bacterial shapes that will be observed in this laboratory exercise.

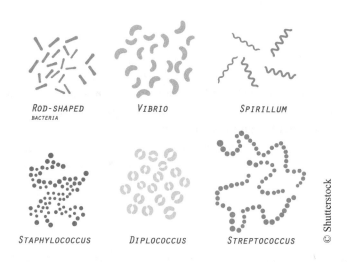

BASIC SHAPES OF BACTERIA

ROD-SHAPED BACTERIA    VIBRIO    SPIRILLUM

STAPHYLOCOCCUS    DIPLOCOCCUS    STREPTOCOCCUS

© Shutterstock

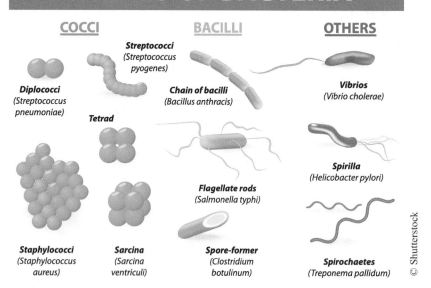

# SHAPES OF BACTERIA

### COCCI    BACILLI    OTHERS

**Diplococci**
(Streptococcus pneumoniae)

**Streptococci**
(Streptococcus pyogenes)

**Tetrad**

**Chain of bacilli**
(Bacillus anthracis)

**Vibrios**
(Vibrio cholerae)

**Spirilla**
(Helicobacter pylori)

**Flagellate rods**
(Salmonella typhi)

**Staphylococci**
(Staphylococcus aureus)

**Sarcina**
(Sarcina ventriculi)

**Spore-former**
(Clostridium botulinum)

**Spirochaetes**
(Treponema pallidum)

© Shutterstock

Types of bacteria. Bacteria are classified into five groups according to their basic shapes: spherical (cocci), rod (bacilli), spiral (spirilla), comma (vibrios), and corkscrew (spirochaetes).

## Gallery of Bacteria Key

Examples of species of bacteria exhibiting the various shapes:
(Note: the bacteria have been stained with a simple stain, unless otherwise noted)

| Gallery/slide # | Shapes | Bacteria |
|---|---|---|
| | **Cocci:** | **Cocci:** |
| 1 | cocci – mixture clusters and tetrads | Staphylococcus aureus and Micrococcus luteus |
| 1 | cocci – clusters and single | Staphylococcus aureus |
| 1 | cocci – tetrads | Micrococcus luteus |
| 2 | cocci – chains pairs and single | Streptococcus species: Enterococcus faecalis Streptococcus pyogenes Streptococcus pneumonia (diplococcic) |
| 3 | cocci – diplococci | Neisseria species: Moraxella catarrhalis Neisseria gonorrheae |
| | **Rods (bacilli)** | **Rods (bacilli)** |
| 4 | rods – single and chains | Bacillus megaterium (with or without endospores) |
| 5 | rods – (coccobacilli) small, single, chains | Escherichia coli |
| 6 | rods – long, single chains | Pseudomonas aeruginosa |
| 7 | rods – pleomorphic (palisades) | Corynebacterium species: Corynebacterium xerosis Corynebacterium diptheriae |
| | **Rods (curved or spiral)** | **Rods (curved or spiral)** |
| 8 | rods – curved or spiral | Rhodospirillum species or Spirillum volutans |
| | **Separate Slides** | **Separate Slides** |
| 9 | **rods – spiral** | typical Spirillum |
| 10 | **bacteria, yeast and blood cells for size comparison** | bacteria, yeast, and blood cells |

*Examining Different Shapes of Bacteria*

1. Use prepared slides to examine the different shapes of bacteria.
2. Make sure you feel confident in recognizing the different shapes, sizes, and arrangements of bacteria.
3. Draw your observations in the Evaluation of Results section.

## EVALUATION OF RESULTS (EXERCISE 5: OBSERVING BACTERIA)

### Purpose
*Data*

Use the circles provided for drawing the organisms seen under oil immersion (100X) from the slides provided.

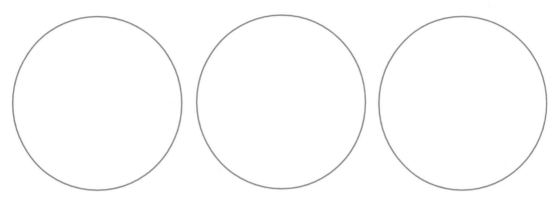

1. _____   2. _____   3. _____

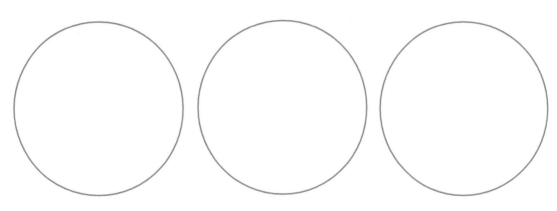

4. _____   5. _____   6. _____

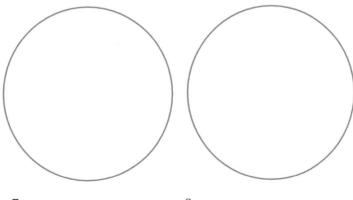

7. _____   8. _____

Data continued:

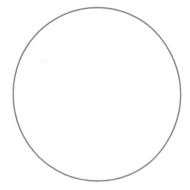 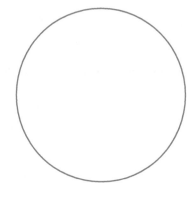

9. _____          10. _____

## CONCLUSIONS, DISCUSSIONS, AND QUESTIONS

1. Describe two of the differences between bacteria (procaryotes) and some of the eucaryotes studied so far.

Prokaryotes          |    eukaryotes
simple               |    nucleus
no organelles        |    organelles

2. List the three major shapes of bacteria, and give examples of each.

cocci - Staphylococci aureus

bailli - bacillus anthracis

~~rods (curved/spiral)~~

vibrios - vibrio cholerae

# Aseptic Technique

## OBJECTIVES

At the conclusion of the exercise, you should...

1. know the purpose of aseptic technique.
2. be able to perform aseptic technique.
3. learn how to use a Bunsen burner properly.
4. learn how to transfer bacteria from test tubes containing broth or agar.
5. learn how to transfer bacteria from Petri plates.
6. learn how to transfer bacteria from broth and agar to microscope slides.
7. be able to use aseptic technique when performing all procedures in a microbiology lab.

## INTRODUCTION

Bacteria must be **cultured** (grown) in order to study them. They are **inoculated** (introduced) into culture media to grow them in order to perform the tests necessary for these studies. The inoculations have to be performed without adding other unwanted microbes (**contaminants**). The process of growing bacteria in **pure** (uncontaminated) cultures is called **aseptic** technique. This exercise will give you the guidelines for performing aseptic technique when working with bacteria, not only for keeping the cultures pure, but for safety reasons. You will use this technique with all the exercises in this manual, and will be referred to this exercise often.

## MATERIALS

Safety: Biosafety Level 1

Cultures:

Broth cultures of bacteria - BSL1
Agar slants of bacteria - BSL1

Media:

Trypticase soy broth
Trypticase agar slant

Supplies:

Inoculating loop
Inoculating needle

## PROCEDURES
Day 1 (Inoculations)

*Technical Background*

Specialized tools are used when handling bacteria. The most commonly used one is the **inoculating loop**. Others are **inoculating needles**, wood sticks, glass rods, cotton swabs, syringes, **pipettes**, and **mechanical pipetters** for the pipettes. The inoculating loop and needle will be described in this exercise. The others will be introduced in later exercises.

The inoculating loop is a wire with a loop on the end, attached to an insulated handle. The needle is a straight wire on an insulated handle. They are both heated and cooled before using to transfer bacteria aseptically. The **Bunsen burner** is used to produce a flame from gas for heating.

*Aseptic Technique: Transferring from a Test Tube to a Test Tube*
(**Note:** Observe the demonstration by your instructor for this procedure.)

1. Turn on the Bunsen burner.
2. Adjust the flame so that there are two color cones in the flame (see diagram).
3. Heat the inoculating loop (or needle) in the inside cone until it is red hot.
4. Cool the loop by holding it in the air for a few minutes or touching a sterile area of the media or container (DO NOT wave it around).
5. Use the test tube with bacteria in a broth first. Vortex or mix the culture to distribute the organisms throughout the broth.
6. Hold the test tube in one hand and remove the cap with the ring and pinkie fingers of the hand holding the inoculating loop.
7. Pass the top of the test tube through the flame.
8. Insert the loop into the liquid.
9. Remove the loopful of liquid.
10. Pass the top of the tube through the flame again, and replace the cap.
11. Be careful with the inoculating loop that has the loopful of bacteria in it. Transfer the loopful of bacteria to the test tube containing sterile broth by removing the cap and swirling the loopful of bacteria in the broth. Replace the cap immediately.
12. Once the transfer is complete, immediately flame the loop until it is red-hot.
13. Let it cool before putting it down.
14. Incubate the broth at 37°C until the next laboratory period.
15. Optional: Inoculate an agar slant by repeating Steps 3-10 and passing the loopful of bacteria back and forth over the surface of the slant.

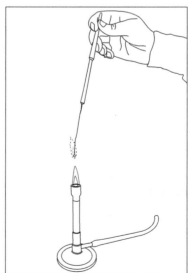

Figure 6.1 Flaming the Loop.

16.    Incubate the agar slant until the next laboratory period.

**Note:** Follow Steps 1 through 10 to transfer from a test tube to a labeled microscope slide or a Petri dish with agar. These aseptic transfers will be described in the following exercises.

## ASEPTIC TECHNIQUE: TRANSFERRING FROM A TEST TUBE

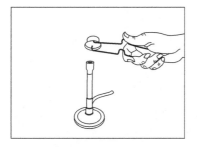

Light Bunsen burner

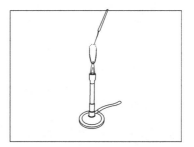

Flame loop

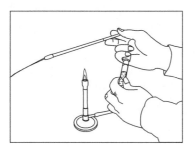

Remove cap

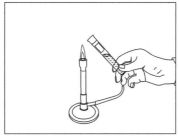

Flame test tube

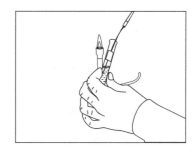

Take loopful of culture

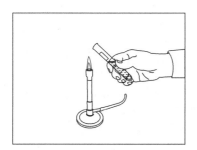

Flame test tube

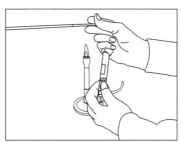

Replace cap

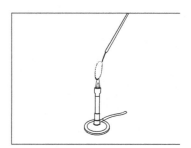

Flame loop

Day 2: Observations

1.  Note the appearance of growth in the broth and agar slant and draw under the Data section.

2.  If no growth is observed, repeat the procedure for the next laboratory period.

## EVALUATION OF RESULTS
## (EXERCISE 6: ASEPTIC TECHNIQUE)

Purpose

Data

Draw the appearance of growth in the broth and agar slant (if applicable)

## CONCLUSIONS, DISCUSSIONS, AND QUESTIONS

1. Discuss two reasons why aseptic technique is used when studying bacteria.

   *uncontaminated*

2. List the advantages for using an inoculating needle compared to using an inoculating loop.

3. List two reasons why there may be no growth observed after inoculating a broth or slant with an organism.

4. List the purposes for using aseptic technique.

5. Define the following:

   Sterile culture media–

   Inoculated culture–

   Pure culture–

   Contaminated culture–

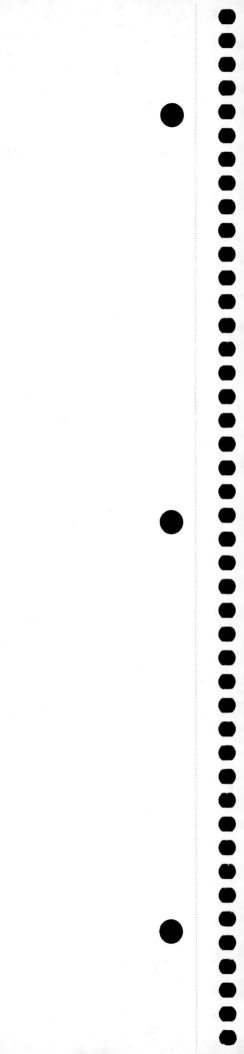

# EXERCISE 7

## Smear Preparation

## OBJECTIVES

At the conclusion of the exercise, you should...

1. be able to make smears from culture broths and agar.
2. understand the difference between broths and agar.
3. understand the different uses of agar slants and plates.
4. understand the different uses of broth cultures.
5. know why the slide is air dried and heat fixed before staining.

## INTRODUCTION

Bacterial cultures are grown in broths (liquid media) and on agar (solid media). Each type of growth has different uses for further studies of bacteria. These uses will be discussed in this exercise and used in later exercises. In this exercise, you will learn how to transfer bacteria from broths and agar to microscope slides to be used for the staining exercises.

## MATERIALS

Safety: Biosafety Level 1

Cultures:

Agar slants, plates, and broths of various bacteria to be used in the staining exercises - BSL1

Supplies:

Microscope slides with frosted ends
Pencil (lead)
Marking pen (China marker, Sharpie, or Gram Stain pen)
Inoculating loop
Inoculating needle
Clothespins (for holding slides)

## Technical Background

Bacteria are grown in broth cultures to yield large numbers of organisms in a small space for further studies. Agar slants are also used to grow bacteria in large numbers for further studies and for transport. Agar slants can be stored for long periods of time. Bacteria are grown on agar plates to observe colony morphology and to check for purity. To make smears, bacteria are air dried on a microscope slide and then heat fixed to kill the bacteria and to adhere the bacteria to the slide so it does not wash off during the staining process.

## PROCEDURES

*Steps to Make a Smear:*

**From broth:**
1. Label a microscope slide on the frosted end using a pencil.
2. Make a dime-sized circle on the underside of the slide (in the center) with the Gram Stain pen (or grease pencil).
3. Follow the aseptic technique described earlier (Ex. #6) for flaming the loop and transferring bacteria from a test tube.
4. Pick up a loopful of bacteria from the broth tube.
5. Transfer the loopful of bacteria onto the slide, within the marked circled area.
6. Spread the bacteria by moving the loop in a circular pattern. Flame the loop!
7. Allow the slide to completely air dry.
8. Heat fix the slide by holding it with a clothespin and passing it quickly through the top of the Bunsen burner flame two or three times.

**From agar (See Figures on next page):**
1. Label another microscope slide on the frosted end with a pencil.
2. Make a circle on the underside of the slide (in the center) with the marking pen, as above.
3. Place a small drop of water on the slide within the circle.
4. Using the needle or the loop, pick up a barely visible amount of bacteria from the agar slant or plate.
5. Mix the bacteria into the drop of water.
6. Spread the bacteria by moving the loop in a circular pattern. Flame the needle or loop!
7. Allow the slide to completely air dry.
8. Heat fix the slide by holding it with a clothespin and passing it quickly through the top of the Bunsen burner flame two or three times.

**Note: Follow the instructions in the staining exercises (#8-10) once you have air dried and heat fixed your smears.**

CAUTION: It is tempting to speed up the drying of the smear by heating the slide over the Bunsen burner flame. <u>Do not do this!</u> Heating bacteria in the presence of water seriously distorts them, and your results will be difficult to interpret. Allow the smears to dry completely (i.e., no sign of moisture) before heat fixing. Once heat fixed, the smears can be stored for later use.

## SMEAR PREPARATION PROCEDURE

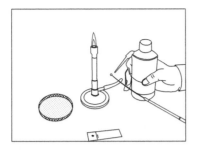

1. After flaming and cooling the loop, add a drop of water to the loop.

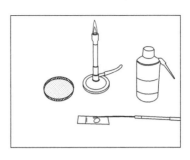

2. Draw a circle on the bottom of the slide, and place the water droplet on the microscope slide in the area of the circle.

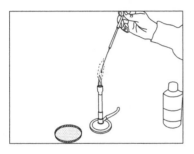

3. Flame and cool the loop.

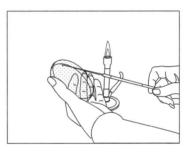

4. Pick up a small amount of culture from an isolated colony on a plate.

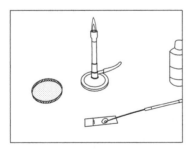

5. Spread the loop of bacteria in the water droplet in the area of the circle.

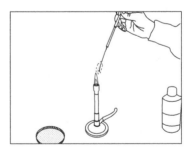

6. Flame and cool the loop.

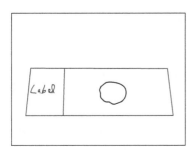

7. Air dry the slide.

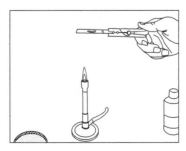

8. Heat fix the slide by passing it over the top of the flame a few times.

## EVALUATION OF RESULTS
## (EXERCISE 7: SMEAR PREPARATION)

Purpose

Data

This exercise demonstrates the first steps for preparing slides to be stained; therefore, data from this exercise will be entered in subsequent staining exercises.

## CONCLUSIONS, DISCUSSIONS, AND QUESTIONS

1. List two reasons for heat fixing the bacteria on a smear.

2. Discuss the advantages and disadvantages of using broths and agar for making smears.

3. Why would the inoculating needle be better than the inoculating loop for making a smear from agar?

# EXERCISE 8

## Simple Staining

### OBJECTIVES

At the conclusion of the exercise, you should...

1. know the advantages of staining bacteria.
2. explain the mechanism of staining.
3. perform a simple stain on the cultured bacteria provided.
4. perform a simple stain on the microorganisms from scrapings around the teeth and gums.
5. observe **pleomorphic** and **palisade** arrangements of bacteria.

**Exercises Required**
Exercise 1: The Brightfield Microscope
Exercise 5: Observing Bacteria
Exercise 6: Aseptic Technique
Exercise 7: Smear Preparation

### INTRODUCTION

Bacteria are very small and transparent when observed with a wet mount preparation. In order to observe their cell characteristics, they need to be stained (dyed). This method consists of preparing a smear that is air dried and heat fixed (as in the smear preparation exercise) and adding a stain to the bacteria on the slide.

### MATERIALS

Safety: Biosafety Level 1

Cultures:

*Corynebacterium xerosis* - BSL1
*Corynebacterium pseudodiptheriticum* - BSL1

Supplies:

Microscope slides with frosted ends
Pencil
Gram Stain marking pen (or China grease marking pen)
Toothpicks (sterile)

Clothespins
Inoculating loop
Inoculating needle
Loeffler's methylene blue
Bibilous (blotting) paper

## Technical Background

Bacteria are very small and transparent when observed with a wet mount preparation. In order to observe their cell characteristics, they need to be stained (dyed). This method consists of preparing a smear that is air dried and heat fixed, and adding a stain to the bacteria on the slide.

**Simple staining** is useful in determining the basic morphology of an organism. Simple staining involves only one reagent. Examples of simple stains are crystal violet, basic fuchsin, and methylene blue. **Pleomorphism** is a morphological characteristic that pertains to an organism's ability to demonstrate several different shapes. The Corynebacteria are rod-shaped, but when grown on certain media, they will appear club-shaped and needle-shaped. The Corynebacteria also exhibit **palisade arrangement**. This type of cell arrangement is described as a "picket fence" arrangement.

Bacteria are usually negatively charged anionic (-). The simple dyes used to stain bacteria have a positive charge cationic (+); therefore, they are attracted to each other. These dye molecules are also strongly colored so that our eyes can see the stained bacteria.

## PROCEDURES

1.  Prepare air dried, heat fixed smears of the *Corynebacterium* species.

2.  Prepare a smear of material from your teeth as follows:
    a.  Put a loopful of water on a clean slide.
    b.  Use a sterile toothpick to scrape some material from between your teeth.
    c.  Smear this material in the water; air dry and heat fix.

3.  Place the slides on the staining rack.
4.  Flood the prepared smears with Loeffler's methylene blue for 1 minute.
5.  Pick up each slide with the clothespin and rinse the dye off with water.
6.  Blot using a piece of bibulous paper; then air dry.
7.  Examine under oil immersion.
8.  Draw your observations in the data section of Evaluation of Results.
9.  (Optional) Make a wet mount of your gum scraping without staining it, and compare it to the stained slide.

## EVALUATION OF RESULTS
## (EXERCISE 8: SIMPLE STAINING)

Purpose

Data

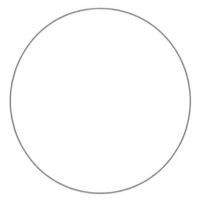

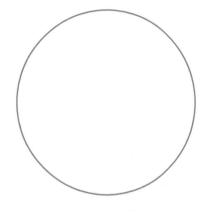

*C. xerosis*                                    *C.  pseudodiphtheriticum*

## CONCLUSIONS, DISCUSSIONS, AND QUESTIONS

1. If you had made a wet mount of the scrapings from your teeth, would you have seen the same organisms as you did with the simple stain? Why or why not?

2. Why is the simple stain used for observing bacteria?

3. What shapes of bacteria were observed from the teeth and gum scrapings?

4. Use your textbook to find the names of 3 genera of bacteria that might be present in the scrapings from your teeth.

# E X E R C I S E 9

## The Gram Stain

## OBJECTIVES

At the conclusion of the exercise, you should...

1. be able to explain the purpose of the Gram stain (differential stain).
2. be able to explain what happens in all the steps of the Gram stain.
3. perform and interpret the Gram stain.

**Exercises Required**
Exercise 1: The Brightfield Microscope
Exercise 6: Aseptic Technique
Exercise 7: Smear Preparation

## INTRODUCTION

The Gram stain was developed in 1884 by a bacteriologist named Christian Gram. The procedure separates bacteria into two major groups: the Gram positives that stain purple and the Gram negatives that stain pink or red. It is important to master the Gram stain because it is the first step in identifying bacteria. You will see this in practice when you identify an unknown organism that you will be given later in the semester. In this exercise, you will Gram stain Gram-positive and Gram-negative bacteria.

## MATERIALS

Safety: Biosafety Level 1

Cultures:

Gram stain practice cultures grown in Trypticase Soy Broths (TSB) and/or Trypticase Soy Agar (TSA) plates and slants

*Escherichia coli* - BSL1
Gram-negative rods (small cocci bacillus rods)

*Staphylococcus epidermidis* - BSL1
Gram-positive cocci (clusters)

## Supplies:

Microscope slides with frosted ends
Pencil
Gram stain pen or grease pen
Wash bottle of tap water
Clothespin
Gram stain reagents:        Crystal violet
                            Gram's iodine
                            Ethyl alcohol
                            Safranin

## Technical Background

The Gram stain is a differential stain because bacteria react differently to the multiple reagents that are used. When the Gram stain is performed correctly, it divides nearly all bacteria into two major groups: Gram-positive or Gram-negative. It is one of the most useful differential stains used in diagnostic microbiology. The first two stains–crystal violet and iodine–form a purple complex within the cell wall. The third step, decolorization with ethanol, is not able to wash the purple complex out of Gram-positive cells because they have thick cell walls. Therefore, the cells retain the original stain and are purple at the end of the procedure, i.e., Gram-positive. Iodine (the mordant) is essential; without iodine, the crystal violet is rapidly removed from Gram-positive cells.

Conversely, Gram-negative bacteria have thin cell walls that cannot retain the crystal violet iodine complex when ethanol is applied. These cells become decolorized. The final stain in the series–Safranin–is applied as a counterstain to render these colorless cells pink, and, therefore, visible. As a result, Gram-negative bacteria are pink or red at the end of the procedure.

If the cells are old or stressed (by, for example, heat or cold), their cell walls become thin, and the crystal violet iodine complex is washed out of all cells, Gram-positive or Gram-negative. Therefore, it is best use fresh cultures (24 hours) for the Gram stain, or Gram-positive organisms may appear Gram-negative.

## PROCEDURES

### Preparing a Gram Stain Slide

See previous directions for making smears from broths and agar.

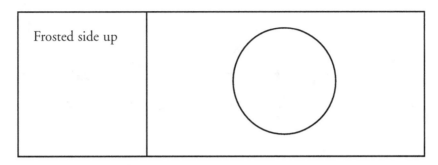

1. Place a loopful of tap water in the center circle of each slide (see diagram).
2. Prepare a different smear from each of the cultures provided.
3. Air dry and heat fix the smears.
4. Place the slide on the staining rack over the sink, and perform the steps of the Gram stain as shown in the diagrams:
   a. Cover (flood) the slide with crystal violet and leave on for 1 minute.
   b. Hold the slide at a 45° angle with a clothespin and gently rinse the slide with water for a few seconds, using the wash bottle.
   c. Cover (flood) the slide with iodine and leave on for 1 minute.
   d. Rinse with tap water, as above.
   e. Hold the slide at 45° with a clothespin and decolorize with 95% ethyl alcohol by dripping on the slide for about 30 seconds* or until no large amounts of purple wash off the slide. This step is critical. [*Note: The degree of decolorizing will depend on the thickness of the smear and the type of decolorizer used. Times may vary by the type of decolorizer used.] Do not over-decolorize!
   f. Rinse with tap water, as above.
   g. Counterstain, covering the slide with safranin for 1 minute.
   h. Rinse with tap water, as above.
   i. Blot dry with bibulous paper and air dry. There is no need to coverslip the slide.
5. Examine the slide under oil immersion (proper observations cannot be made at lower magnification).
6. Draw and record your observations of the Gram stain reactions of the cultures provided on the Data/Results section, i.e., short red rods, large purple cocci, etc. Label the organisms accordingly.
7. Interpret your results: Always include the Gram reaction, shape, and arrangement for the organism, i.e., Gram-negative rods arranged singly, Gram-positive cocci in clusters, etc.
8. Compare your result for each organism with the expected result given.
9. Use the Gram stain troubleshooting guide in the Appendix to explain any discrepancies.

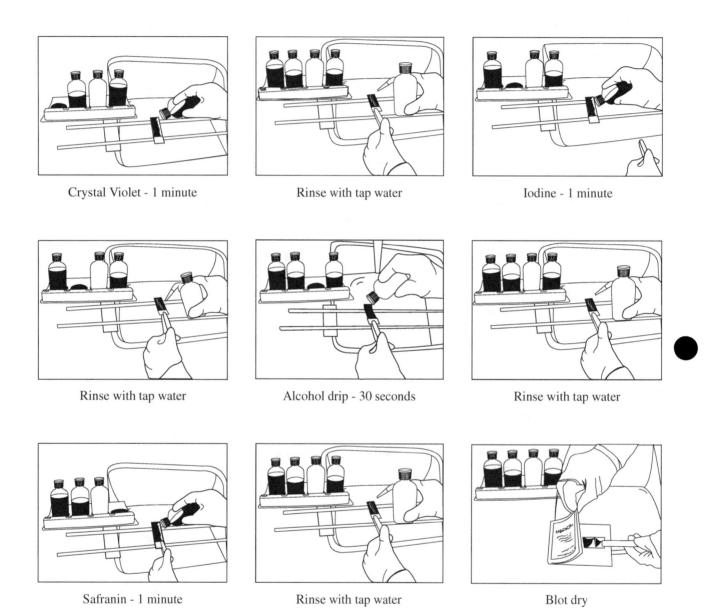

Figure 9.1 Gram Stain Procedure.

| Table 9.1 | Gram Stain Procedure | | |
| --- | --- | --- | --- |
| **REAGENTS** | **TIME APPLIED** | **REACTIONS** | **APPEARANCE** |
| UNSTAINED SMEAR | | | CELLS ARE COLORLESS AND DIFFICULT TO SEE. |
| CRYSTAL VIOLET | 1 MINUTE, THEN RINSE WITH WATER | BASIC DYE ATTACHES TO NEGATIVELY CHARGED GROUPS IN THE CELL WALL, MEMBRANE, AND CYTOPLASM. | BOTH GRAM-NEGATIVE AND GRAM-POSITIVE CELLS ARE DEEP VIOLET. |
| GRAM'S IODINE (MORDANT) | 1 MINUTE, THEN RINSE WITH WATER | IODINE STRENGTHENS THE ATTACHMENT OF CRYSTAL VIOLET TO THE NEGATIVELY CHARGED GROUPS. | BOTH GRAM-NEGATIVE AND GRAM-POSITIVE CELLS REMAIN DEEP VIOLET. |
| ALCOHOL OR ACETONE-ALCOHOL MIX (DECOLORIZER) | 10 TO 15 SECONDS, THEN RINSE WITH WATER | DECOLORIZER LEACHES THE CRYSTAL VIOLET AND IODINE FROM THE CELLS. THE COLOR DIFFUSES OUT OF GRAM-POSITIVE CELLS MORE SLOWLY THAN OUT OF GRAM-NEGATIVE CELLS BECAUSE OF THE CHEMICAL COMPOSITION AND THICKNESS OF THE GRAM-POSITIVE CELL WALLS. | GRAM-POSITIVE CELLS REMAIN DEEP VIOLET, BUT GRAM-NEGATIVE CELLS BECOME COLORLESS AND DIFFICULT TO SEE. |
| SAFRANIN (COUNTERSTAIN) | 1 MINUTE; THEN RINSE THOROUGHLY, BLOT DRY, AND OBSERVE UNDER OIL IMMERSION | BASIC DYE ATTACHES TO NEGATIVELY CHARGED GROUPS IN BOTH CELL TYPES. FEW NEGATIVE GROUPS ARE FREE OF CRYSTAL VIOLET IN GRAM-POSITIVE CELL, WHEREAS MOST NEGATIVE GROUPS ARE FREE IN GRAM-NEGATIVE BACTERIA. CONSEQUENTLY, GRAM-POSITIVE BACTERIA REMAIN DEEP VIOLET, WHEREAS GRAM-NEGATIVE BACTERIA BECOME PINK OR RED. | GRAM-POSITIVE CELLS REMAIN DEEP VIOLET, WHEREAS GRAM-NEGATIVE CELLS ARE STAINED PINK OR RED. |

Gram stain of *Staphylococcus aureus,* a gram-positive coccus (×252). *Daniel Lim.*

Gram stain of *Neisseria flavescens,* a gram-negative coccus (×252). *CDC.*

## EVALUATION OF RESULTS
## EXERCISE 9: THE GRAM STAIN

Purpose

Data

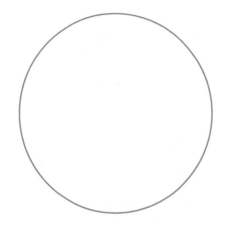

Name of organism: _____

Observations: _____

Interpretation:_____

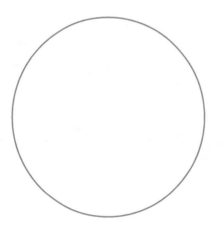

Name of organism: _____

Observations: _____

Interpretation:_____

## CONCLUSIONS, DISCUSSIONS, AND QUESTIONS

1. **Circle** the answer that describes how the bacteria would appear after staining:

   a. If the mordant (iodine) was left out of the procedure.

      Gram-positive bacteria: **pink/purple**
      Gram-negative bacteria: **pink/purple**

   b. If the alcohol was skipped.

      Gram-positive bacteria: **pink/purple**
      Gram-negative bacteria: **pink/purple**

   c. If the cultures were 3 days old.

      Gram-positive bacteria: **pink/purple/both pink and purple cells**
      Gram negative bacteria: **pink/purple**

2. You observe a Gram-stained slide that has pink cocci and purple cocci. List the possible reasons for this result.

3. List the steps of the Gram stain, starting with the aseptic transfer of the culture.

# EXERCISE 10

## Miscellaneous Staining

## OBJECTIVES

At the conclusion of the exercise, you should...

1. know what causes some bacteria to be "acid fast."
2. know the purpose of doing the acid-fast stain.
3. know what bacteria need to be stained using the acid-fast technique.
4. perform the acid-fast stain.
5. observe *Mycobacterium* species that have been acid-fast stained.
6. understand what is an endospore.
7. know what and why bacteria form endospores.
8. be able to perform the endospore stain.
9. learn the characteristics of Gram-positive, endospore-forming rods.
10. observe slides of *Bacillus* species with endospores.
11. know what and why bacteria form capsules.
12. observe slides of bacteria that have been stained with a capsule stain.
13. understand what is a negative stain.

### Exercises Required
Exercise 1: The Brightfield Microscope
Exercise 6: Aseptic Technique
Exercise 7: Smear Preparation

## INTRODUCTION

Bacteria are stained in order to see them easily (e.g., the simple stain) and to aid in their identification (e.g., the Gram stain, a differential stain in which more than one reagent is used and different bacteria react differently). A third use for staining bacteria would be to observe specific structures for identification purposes. Structural stains are used to stain specific structures. The endospore and capsule stains are examples of structural stains. The acid-fast stain is another example of a differential stain (using more than one reagent and the bacteria reacting differently to the reagents). In this exercise, you will perform and observe the acid-fast stain, the endospore stain, and observe a demonstration slide of the capsule stain.

## MATERIALS

Safety: Biosafety Level 1

Cultures:

*Mycobacterium smegmatis* (acid-fast stain) - BSL1
*Staphylococcus epidermidis* (acid-fast stain) - BSL1
*Bacillus megaterium* (endospore stain) - BSL1
*Bacillus subtilis* (endospore stain) - BSL1

Supplies:

Prepared slides showing:

1. Positive AFB (Acid-Fast Bacilli) sputum smears.
2. Endospore stains of *Bacillus* species.
3. Capsule stains of *Klebsiella pneumoniae*. Maneval's capsule stain.

## A. ACID-FAST STAIN (INTRODUCING THE GENUS *MYCOBACTERIUM*)

## PROCEDURES

### Technical Background

The acid-fast stain is used most commonly to aid in the diagnosis of tuberculosis (TB), an infection caused by *Mycobacterium tuberculosis*. Mycobacteria can infect almost any tissue or organ of the body. Mycobacteria have a waxy substance called mycolic acid as part of the cell wall. This waxy substance makes the organism very slow growing and thus difficult to isolate and identify. Mycobacteria can be recovered optimally from clinical specimens when methods are used both to release them from body fluids and cells (called digestion) and to remove or sufficiently reduce competing organisms (called decontamination).

Specimens that are submitted for mycobacterial culture are of two categories: specimens of pulmonary origin (primarily to isolate M. tuberculosis) and specimens of extrapulmonary origin. The specimens of pulmonary origin (usually sputum) are almost always contaminated with normal flora. Because they can contain large numbers of normal contaminating flora and the normal consistency can trap the mycobacteria, they need to be decontaminated and digested, a process which also concentrates the bacilli.

After the digestion and decontamination procedures, the specimen is cultured and smears are prepared for staining. An acid-fast stain is performed, which uses acid-alcohol as the decolorizer. Because these organisms contain mycolic acid in their cell wall, they are resistant to decolorization and retain the original stain. The visualization of acid-fast bacilli in sputum or other clinical material should be considered only presumptive evidence of tuberculosis, since the stain does not specifically identify Mycobacterium tuberculosis. Culturing these bacteria allows for their identification.

*Performing the Steps of the Acid-fast Stain*

1. Using aseptic technique, prepare a slide of *Mycobacterium smegmatis* mixed with a loop-ful of *S. aureus*.
2. Air dry and then heat fix the smear.
3. Flood the slide with Kinyoun's TB Carbolfuchsin (KF) for 4 minutes.
4. Wash the slide gently with water.
5. Drip acid-alcohol on the slide until no more color drains from the smear.
6. Wash gently with water.
7. Counterstain with methylene blue for 30 seconds.
8. Wash gently with water.
9. Blot dry very gently.
10. Air dry.
11. Observe the slide under oil and compare to the figure in the Atlas manual.
12. Interpretation: The acid-fast *Mycobacterium smegmatis* should be rod shaped and appear dark pink to bright red. The *S. aureus* should be cocci and appear blue.
13. Observe the positive AFB sputum smear. Typical acid-fast bacilli (AFB) appear as purple to red, slightly curved, short or long rods (2-8 μm) against the tissue, which stains with the blue counterstain. Note the different appearances between your slide and the stained sputum slide.

## B. ENDOSPORE STAINS (INTRODUCING THE GENUS *BACILLUS*)

## PROCEDURES

### Technical Background

During extreme environmental conditions, such as heat or drought, certain bacteria are capable of forming a specialized cell structure called an **endospore**. This **spore** is unique to certain genera of bacteria. Two of these genera are ***Clostridium*** and ***Bacillus***. Endospores are highly durable, dehydrated cells with thick cell walls. They can survive extreme heat, lack of water, and many toxic chemicals and radiation. Because of their nearly impenetrable cell walls, the Gram-stain method will not stain endospores; therefore, specialized staining methods are necessary. Most of the methods utilize heat to drive the stain into the endospore cell wall. One method, called the Schaeffer-Fulton Method, uses malachite green stain with heat and safranin for a counterstain. The endospore will stain green and the surrounding vegetative cell will stain pink.

There is a cold endospore staining method that utilizes malachite green by flooding the slide for at least 10-15 minutes and then counterstaining with safranin. The endospores will appear a lighter green than when heat is used. This method is safer and cleaner to use.

*Performing the Steps of a Cold Endospore Stain*

1. Prepare a slide of *Bacillus* megaterium or a slide of *Bacillus subtilis*.
2. Air dry and heat fix the slides.
3. Flood the slide with malachite green for 10-15 minutes.
4. Rinse the slide with water.
5. Counterstain with safranin for 30 seconds.

6.  Rinse with water and blot dry.
7.  Air dry the slide.
8.  Examine the slides with oil immersion.
9.  Also, observe the prepared slides of *Bacillus megaterium* endospores.
10. Draw and label your observations in the evaluation portion of the Data section.

## C. CAPSULE STAINS

## PROCEDURES

### Technical Background

Many bacteria produce a slimy layer that surrounds and adheres to the cell. When the slime is loosely bound to the bacterium, it is called a **slime layer**. When highly symmetrical and organized, this layer is called a **capsule**. Capsules play a role in the **virulence (disease-causing ability)** of some bacteria. Capsules are composed of polysaccharides and are water-soluble. Simple stains will not adhere to the capsule; to visualize them, the background and bacteria must be stained, while the capsule remains unstained. This process is called **negative staining**. One method of negative staining uses nigrosine (or India ink), which stains the background black. Crystal violet is used as a counterstain to stain the bacterial cell, thus making the capsule visible as a **clear halo** around the cell. Another method (Maneval's) uses Congo red for the background, where the bacterial cells are stained reddish-brown and the capsules are unstained.

### Performing the Steps of a Capsule Stain

1.  Place a drip of Congo red on a slide.
2.  Transfer some of the bacteria to be stained from a milk culture to the drop of Congo red on the slide and mix.
3.  Spread the mixture of Congo red and bacteria over the slide to about the size of a quarter.
4.  Allow the slide to air dry. DO NOT HEAT FIX THE SMEAR.
5.  Flood the slide with the Maneval's stain (second solution) for 1 minute.
6.  Rinse the slide with water.
7.  Air dry.
8.  Examine under oil immersion. The capsules appear unstained, the background is purple-red, and the bacterial cells are stained red to reddish-brown, depending on the thickness of the smear.
9.  Observe the prepared slides of Klebsiella pneumoniae capsules.
10. Draw and label your observations in the evaluation portion of the Data section.

# EVALUATION OF RESULTS
# EXERCISE 10: MISCELLANEOUS STAINS

Purpose

Data

**A. Acid-fast Stain**                                        **Prepared Slides**

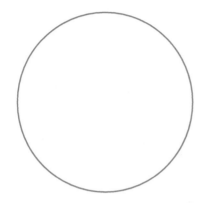
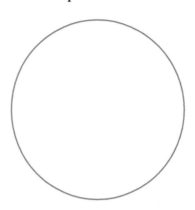

   *M. smegmatis + S. epidermidis*      Acid-fast

**B. Endospore Stain**

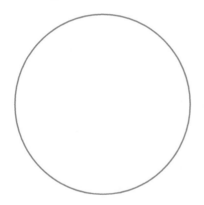
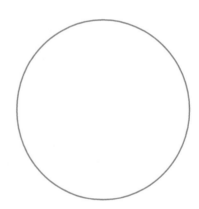

   *Bacillus megaterium*  or  *Bacillus subtilis*

**C. Capsule Stain**

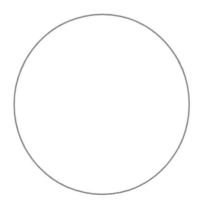

   *K. pneumoniae* – prepared slide (BSL2)

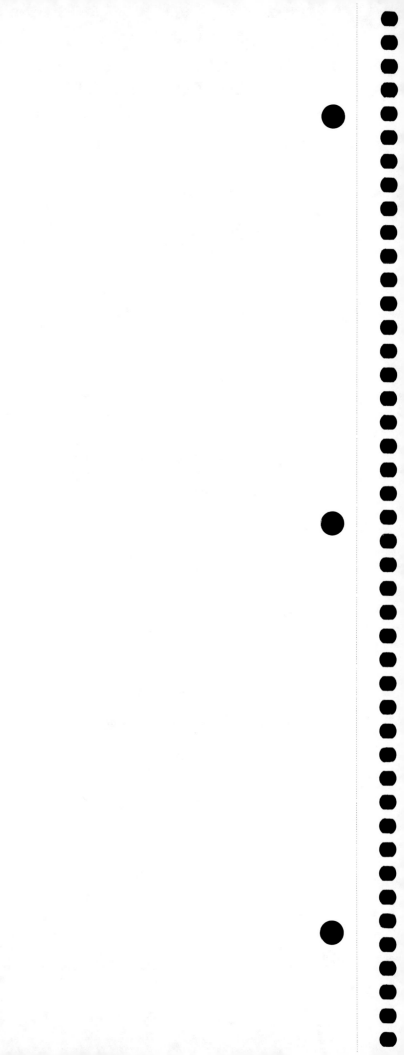

# CONCLUSIONS, DISCUSSIONS, AND QUESTIONS

1. What is the decolorizing agent used in the acid-fast stain? How is this different from the decolorizing agent used in the Gram stain?

2. What diseases are diagnosed by using the acid-fast stain?

3. Why don't endospores stain with the Gram stain method?

4. Discuss some differences between the genus *Bacillus* and *Clostridium*, including their oxygen requirements, pathogenicity, and endospore characteristics.

5. Explain how capsules contribute to the pathogenicity of some bacteria.

6. A postal worker shows his doctor a large, black, crusty sore on his hand. A Gram stain of the oozing liquid shows large Gram-positive rods in chains. Some of the rods are pink and have clear, oval shapes inside. What structural stain should be done immediately on a fresh smear of the oozing liquid? What is the reason for your answer?

# EXERCISE *11*

## Culture Media Preparation

### OBJECTIVES

At the conclusion of the exercise, you should...

1. understand the uses for culture media.
2. know how to make a general type of culture medium.
3. know how to perform the calculations necessary to make different amounts of culture media.
4. be able to differentiate special types of media.
5. understand how autoclaving results in sterilization.
6. learn how to keep contaminates out of sterile media.

### INTRODUCTION

Bacteria need nutrition to grow. Media (*medium*, singular) is the food that is used for culturing bacteria and other microorganisms. It can be in liquid, solid, or semisolid form. Media must contain water, carbon, nitrogen, minerals, and growth factors. In addition, media must be the right pH, as well as sterile. In this exercise, you will prepare a general-purpose type of medium: nutrient agar plates. You will also be introduced to some of the other media that you will be working with. Refer to Appendix J (Selected Media Descriptions) as new media is introduced.

### MATERIALS

Safety: Biosafety Level 1

Demonstrations:

TSA w/ *E. coli* (BSL1)
TSA w/ *S. epidermidis* (BSL1)

Media:

Nutrient agar powder

Supplies:

Weighing balance (digital)
Weighing paper (or dishes)
Tongue depressor (or spatula)
100 mL graduate cylinder
250 mL flask
Stirring rod
5 sterile Petri plates
RO (reverse osmosis) water

## PROCEDURES
Day 1

*Technical background*

Nutrient agar is a **general-purpose** medium, consisting of beef extract, peptone, and agar. Difco is a company that provides pre-mixed, **dehydrated** culture media. Before media-making companies existed, the microbiologist had to weigh out the different ingredients.

**Selective** media will allow only certain types of bacteria to grow. They usually have inhibitory substances that restrict the growth of other unwanted bacteria. An example of this is Columbia (CNA) media, which restricts the growth of some Gram-negative rods, while permitting Gram-positive bacteria, such as staphylococci and streptococci, to grow.

**Differential** media contains various substances that cause some bacteria to take on a different appearance from other species. **EMB** (eosin methylene blue) agar is an example of a differential (and selective) medium used for the isolation of a specific group of Gram-negative rods, called the enteric bacteria. It contains the milk sugar lactose and allows organisms to be differentiated by their ability to ferment lactose.

**Sterilization** of culture media is done to eliminate contaminating microorganisms from the environment. Complete sterilization occurs at 250°F (121.6°C) at 15 pounds per square inch (psi) of steam pressure. Autoclaves provide this type of sterilization.

*Preparing Culture Media*
Work as a table, assigning the different tasks, as needed.

1. Weigh out the appropriate amount of nutrient agar for making 100 mL. This amount is written on the side of the bottle, along with the ingredients. (You will have to calculate the correct number of grams to weigh, based on the instructions on the side of the bottle of media. In this case, nutrient agar is 23 grams per liter. You will have to set up a ratio to calculate the required amount to make 100 mL.
2. Place the powder into a 250-mL flask and add 100 mL water.
3. Swirl the flask until the media is hydrated. (Note: The only time it is necessary to boil the agar to dissolve it is if it is being dispensed in tubes and then autoclaved.)
4. Adjustment of the pH usually is not necessary. (FYI: It should be around 7.0.)
5. Label the flask.
6. Sterilize the flasks in the autoclave.

7. Turn on two water baths to 50°C.
8. Place the sterilized flasks in the 50°C water baths.
9. Set up 5 sterile Petri plates on your lab bench.
10. When the flask is cool enough to touch, pour the plates.
11. Pour about 20 mL into each of 5 sterile Petri plates. This can be judged by pouring enough to cover the bottom of the plate, plus a little more.
12. Allow the plates to cool on your bench until the agar has solidified.
13. Students should save plates in their lab locker. They will be used later. Store all plates closed and upside down!
14. Clean up! Dirty and unused glassware should go into the decontamination tray. Weighing paper should go in the regular trash. Caps should be put back on the jars of nutrient agar.

## PROCEDURES
Day 2

*Preparing Culture Media, Cont'd*

1. Examine prepared plates.
2. Use prepared plates for streaking cultures in Exercise 12.

## EVALUATION OF RESULTS
## EXERCISE 11: CULTURE MEDIA PREPARATION

Purpose

Data

Show your calculations for the amount of agar you weighed to make 100 ml of nutrient agar.

Describe the surface of the agar on the opened plate and on the unopened plate (number of colonies, including color, size, and shape). Record the appearance and interpretation of the Gram stain for the colonies on the opened plate:

| Colony | Number of this type | Color | Size | Shape | Gram stain |
|--------|--------------------|-------|------|-------|------------|
| 1 | | | | | |
| 2 | | | | | |
| 3 | | | | | |

## CONCLUSIONS, DISCUSSIONS, AND QUESTIONS

1.  If you were given a bottle of nutrient broth powder (8 g/l) and a bottle of agar, how much agar would be needed to make 100 mL? (Note: A broth can be made into agar by adding enough agar to make a 1.5% (w/v) solution, or 15 g/l).

2.  Why is agar a good ingredient for converting liquid media to solid media?

3.  Explain what selective media is and how it is used.

4.  Describe three other types of sterilization techniques that are used in the microbiology laboratory.

5.  What is a "contaminant"?

# E X E R C I S E 12

## Streak Plate Technique and Colony Morphology

### OBJECTIVES

At the conclusion of the exercise, you should...

1. be able to perform the T streak plate method for isolating bacteria.
2. learn the Quad streak plate method for isolating bacteria.
3. be able to streak bacteria for isolation from a mixture of bacteria.
4. list other methods for obtaining pure bacterial cultures.
5. recognize and describe different colony types of bacteria.
6. learn the proper way to label Petri plates.

### INTRODUCTION

Microbes exist in mixed populations in nature. In order to study a given organism, it must be in a **pure culture** (containing one kind of microbe). This exercise will demonstrate how to isolate bacteria from a mixed culture, using the **T streak plate method**.

### MATERIALS

Safety: Biosafety Level 1

Cultures:

TSB mixtures of the following pair of bacteria:

*E. coli* (BSL1) and *S. epidermidis* (BSL1)

Media:

TSA plates

## PROCEDURES
Day 1 (First Isolation)

### *Technical Background*

Isolation of bacteria by **dilution** techniques is necessary in order to obtain a pure culture for further studies. The streak plate is the most-used dilution technique. When a loopful of bacteria is streaked across the agar plate, the bacteria are distributed across the surface of the agar. The more streaks, the more the bacteria will be diluted until, in theory, only one cell is left to grow and give rise to a colony of the same bacteria. There are different types of streak patterns, the **T streak** method and the Quad streaking method. Other dilution methods include the **spread plate technique** and the **pour plate technique**. These will be introduced in later exercises.

**Colony morphology** will help determine whether a pure culture was obtained, since bacteria of the same species will produce nearly identical colonies. **Colony characteristics** should include color, size, shape, margin, and consistency (or texture). Use the following guidelines for describing colonies of bacteria. (Note: It is best to choose colonies that are isolated and not too crowded.) Some of these descriptions will have variations and combinations. For example, a colony could be described as very large, dark brown, slightly raised, slightly spreading, very mucoid.

Size: pinpoint, small, medium, large
Color: white, cream, tan, brown, black, purple, yellow, red, etc.
Shape (elevation): raised, flat, convex, dimpled (caved in), fried egg, growth into agar
Margin (edge): smooth, spreading (swarming), irregular, wavy, rhizoid, filamentous
Consistency (texture): shiny, glistening, dry, powdery, wrinkled, rough, dull or matte, mucoid

**COLONIAL MORPHOLOGY**

**Margin**

smooth          irregular          filamentous          rhizoid          wavy/swarming

**Margin continued**

entire          filamentous          spreading

**Shape**

raised          flat          convex          fried egg          growth into medium

The mixture used for this exercise has two of the three different species of bacteria listed below. They each have characteristic colony morphology. See demonstrations provided.

*S. epi* — medium-sized, white, slightly raised, smooth, shiny to dull colonies
*E. coli* — medium-sized, grey, raised, smooth, shiny colonies

*Performing a Streak Plate Technique*

Observe the instructor's demonstration before you begin.

1. Label a TSA Petri plate on the bottom half around the edge with a Sharpie marking pen. The label should include any or all of the following: **Name**, **Date**, **Class**, **Section #**, **Exercise #**, and/or a short name of the exercise. One example for labeling is: J. Doe 6/8/98  Bio 210, #1,  ex. 12, mix.  This may vary with different schools or instructors.

2. Draw a "T" on the bottom of the plate to delineate the 3 areas for dilution of the organism (see T streak diagram below).

3. Follow aseptic technique for flaming and cooling the loop (as described in the aseptic technique exercise).

4. Mix the broth culture by gently swirling the test tube.

5. Obtain a loopful of the mixture (this will be the only time you will obtain the specimen).

6. Turn the Petri plate over, onto its top.

7. Pick up the bottom of the Petri plate with one hand, leaving the lid on the lab bench.

8. Starting  at the edge of the agar, streak the **Primary inoculation area** (above the crossbar of the "T"). Hold the loop like you would hold a pencil and gently touch the surface of the agar. Be careful not to gouge it. Make as many streaks as possible, covering as much of the area without overlapping previous streaks.

9. Flame and cool the loop.

10. Turn the plate about a quarter turn.

11. Streak the **Secondary inoculation area** by going back into the primary area with your loop only once or twice. Streak this area in the same manner as the primary area, covering as much agar as possible without overlapping the previous streaks.

12. Flame and cool the loop.

13. Turn the plate another quarter turn.

14. Continue streaking as above, passing into the secondary area once or twice and then streaking out to the end of the third inoculation area. (See diagrams.)

15. Incubate the plate **upside down** at 35°C until the next lab period.

**Streaking a plate with the T streak technique**

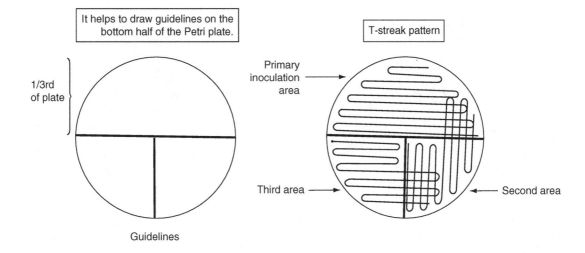

Streaking a Plate with the T Streak Technique.

Streaking a plate with the Quad streak technique

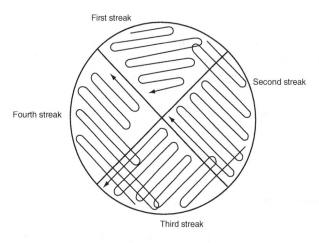

## STREAK PLATE PROCEDURE (T STREAK)

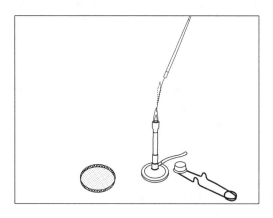

After flaming and cooling the loop, pick up a loopful of the culture or a portion of a colony from a fresh plate, then..............

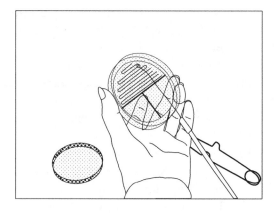

Starting on one side, streak the primary area by streaking about 1/3 of the plate. Flame and cool the loop..............

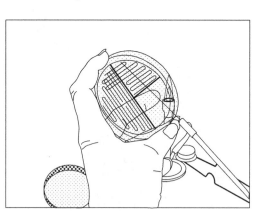

Turn the Petri plate about 1/4 turn and streak the secondary area by going back into the primary area only once or twice and then streaking four to five times in the secondary area. Flame and cool the loop.........

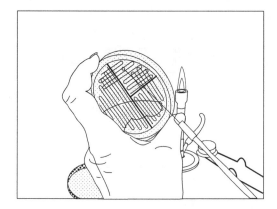

Turn the plate another quarter turn and continue streaking by passing into the secondary area once or twice and then streaking out to the end of the plate.

## PROCEDURES
Day 2 (Second Isolation)

*Describing Colony Morphology*

1. Examine the streak plate for the different colony types.
2. Circle and number the different types of colonies.
3. Describe the colony morphology using the guidelines listed in Day 1 procedures.
4. If you did not obtain two distinct colonies types in the third area of inoculation, restreak for isolation from mixed areas on the plate. Do this by taking a sample from the primary area and follow the T streak procedure using a new TSA plate.
5. If there are two distinct isolated colonies on the plate, restreak each colony type onto a new TSA plate. For each, take a small sample by touching the top of an isolated colony with a needle or a loop and transferring to the primary area of a new TSA plate. Streak for isolation, as instructed previously.
6. Incubate the TSA plates at 35°C until the next lab.
7. Reincubate the original streak plates until the next lab period for more growth, if necessary.
8. Save all streak plates in the refrigerator until the exercise is completed.

## PROCEDURES
Day 3 (Final Evaluation)

*Evaluating Your Results*

1. Examine all of the streak plates for isolated colonies.
2. Compare the results with the original streak plate with the mixture of the bacteria.
3. If necessary, continue practicing streaking for isolated colonies of each of the colony types for up to a total of 2 plates after the initial mixture plate.

## EVALUATION OF RESULTS
## EXERCISE 12: STREAK PLATE TECHNIQUE AND COLONY MORPHOLOGY

Purpose

Data

| Colony # | Color | Size | Shape (elevation) | Margin (edge) | Consistency | Gram Stain |
|----------|-------|------|-------------------|---------------|-------------|------------|
|          |       |      |                   |               |             |            |
|          |       |      |                   |               |             |            |

## CONCLUSIONS, DISCUSSIONS, AND QUESTIONS

1. Why are Petri plate cultures incubated upside down?

2. Discuss reasons why only one colony type would appear after incubation, even though the original broth contained two different species of organisms.

3. What is the purpose of flaming the loop between the streak areas?

4. How could you tell if one of the colonies on the plate was a contaminate that fell in from the air?

# E X E R C I S E  *13*

## Specimen Transport & Ubiquity of Microorganisms

### OBJECTIVES

At the conclusion of the exercise, you should...

1. be able to collect a sample with a specimen transport swab.
2. know why specimen transport devices and media are used.
3. understand the purpose of RODAC plates.
4. observe how ubiquitous microbes are.
5. be able to discuss some of the microorganisms found in the environment.
6. perform and discuss the use of quality control in evaluating Gram stains

### INTRODUCTION

Quality Control (QC): As you have learned, one of the most important techniques used in microbiology is the Gram stain. In order to be sure that the Gram stain is performed correctly, you must include quality-control slides each time you stain an organism of unknown identity. Quality control is used for many techniques in the laboratory and in industry to make sure all reagents are working correctly and the techniques are performed accurately. You will prepare your own set of quality-control slides for Gram stains by following the technique described below.

### MATERIALS

Safety: Biosafety Level 1

Cultures:

*Pseudomonas stutzeri* (Gram-negative rod) - BSL1
*Micrococcus luteus* (Gram-positive coccus) - BSL1

Supplies:

Slides
Gram stain reagents

## PROCEDURES:

Day 1

1. Use the known Gram-negative and Gram-positive bacterial cultures provided: *P. stutzeri* and *M. luteus.*
2. Label the frosted end of the slide (with the pencil) appropriately (see diagram below).
3. Draw two circles on the non-frosted side of the slide (underneath) with a Gram stain pen (see diagram below).
4. One area is used for the two quality control organisms mixed together.
5. Follow the proper technique for smear preparation, as described in Exercise 7; aseptically add a loopful of each QC organism to this area, mix well, and allow to air dry.
6. Aseptically add a loopful of the unknown culture to the second circle (see diagram below).
7. Prepare at least 6 slides (up to 12) with the QC mixture and store for later use.
8. After air drying, heat fix one at a time immediately before staining.
9. Always observe your QC area first to be sure the results are reliable before evaluating the Gram stain of your unknown organism.

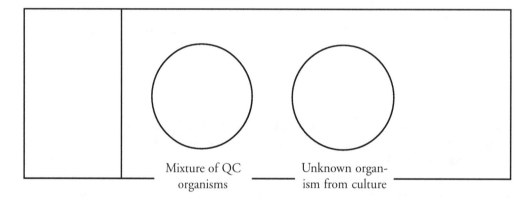

Mixture of QC          Unknown organ-
organisms              ism from culture

## INTRODUCTION

Microorganisms are ubiquitous in nearly every area of the environment. Microbiologists have several tools at their disposal to collect and transport samples to the laboratory to quantify and identify them. In this exercise, you will be introduced to two of these tools and practice the proper collection and evaluation of microbiological samples.

Materials

Media:

1 TSA plate

Supplies:

1 Transport swab
1 Rodac plate

## A. SWAB

## PROCEDURES
Day 1 (Collection)

*Technical Background*

In order to perform laboratory tests on specimens that are collected outside the laboratory, they must be collected and transported properly. It is essential that the container bearing a specimen does not contribute its own **microbial flora**. Also, the original flora should neither multiply nor decrease because of prolonged standing on a hospital ward or in the field.

A variety of containers have been devised for collecting bacteriologic specimens. The most commonly used is a cotton- or Dacron-tipped applicator stick. These must be sterile and remain sterile before specimen collection. One approach uses a sterile disposable culture unit (Culturette, Becton-Dickinson Microbiology Systems), consisting of a plastic tube containing a sterile polyester-tipped **swab** and a small glass ampule of holding medium. This medium maintains a favorable pH and prevents both dehydration of secretions during transport and oxidation and enzymatic self-destruction of any pathogens present. There are other versions of this type of transport media, such as a gel in the bottom of the tube, or a separate test tube into which the specimen swab is inserted.

The unit is removed from its sterile envelope, and the swab is used to collect the specimen. It is then returned to the tube, the ampule is crushed (if there is one), and the swab is forced into the released holding medium. This will provide sufficient moisture for storage up to 72 hours at room temperature. After the specimen arrives in the laboratory, the swab can be removed and used to inoculate the appropriate media.

A variety of transport media is available for prolonging the survival of microorganisms when a significant delay occurs between collection and culturing. Special media is needed for specific types of specimens. One example is for anaerobic specimen transport.

*Collecting and Transporting a Specimen*

1. Follow the directions on the package to open and use the swab.
2. Use the specimen swab to collect a specimen from somewhere away from the laboratory. Some places to collect samples from home include the following: kitchen, bathroom, bed linens, washer, dryer. Some items that could be sampled include toothbrush bristles, hair brush, cutting board, sink handle, TV remote. Be creative. Because of the danger from highly resistant organisms, if you work or volunteer in a health care setting, please do not collect a sample from this location.
3. Follow directions on the package for transporting the swab back to the laboratory.

## PROCEDURES
Day 2 (Inoculation)

*Preparing a Streak Plate of*
*Your Specimen*

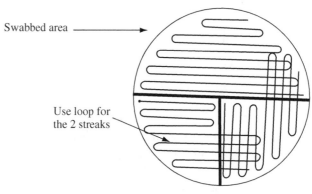

Swabbed area

Use loop for
the 2 streaks

1. Use the specimen swab to
   streak back and forth onto
   one third of the TSA plate
   (see diagram).
2. Flame and cool a loop to
   streak for isolation from
   the primary streak area. Streak the secondary inoculation area by going back into the primary area with your loop only once or twice. Streak this area, covering as much agar as
   possible without overlapping the previous streaks. Flame and cool your loop.
3. Continue streaking the third inoculation by passing into the secondary area once or twice
   and then streaking out to the end of the third inoculation area. Flame your loop.
4. Incubate the TSA plate upside down at 30°C until the next lab period.

## PROCEDURES
Day 3 (Evaluation)

*Evaluating Your Results*

1. If there is a lot of fungus growing on the plate, DO NOT OPEN IT.
2. Use the stereoscope to observe the colonies.
3. Count the total number of colonies obtained and the number of different appearing
   colonies.
4. Describe the colonies in the table in the evaluation portion of the Data section. Repeat
   for your lab partner's sample.
5. Record your results in the evaluation portion of the Data section.

## B. RODAC PLATES

## PROCEDURES
Day 1 (Collection and Inoculation)

*Technical Background*

**RODAC** = Replicate Organism Detection and Counting

RODAC plates are used for the detection and enumeration of microorganisms present on surfaces of sanitary importance. The plates are specially constructed so that the agar medium can be over-filled, producing a meniscus or dome-shaped surface that can be pressed onto a surface for sampling its microbial content. After touching the surface to be sampled with the medium, the dish is covered and incubated and kept at room temperature until it is brought to the lab. It then can be incubated at 30°C or 35°C, depending on the organisms that are being looked for. The presence and number of microorganisms is detected by the appearance of colonies on the surface of the agar. Collection of samples before and after cleaning and treatment with a disinfectant permits the evaluation of the efficacy of sanitary procedures. We will use these as another example of transporting specimens to the lab and for demonstrating the ubiquity of microorganisms. Assigned students will take one plate home and use it to touch an area of their choice—either with or without cleaning. Students are encouraged to be creative when they decide to touch the plate. Some examples of places to touch include the refrigerator, kitchen counter, bathroom sink, bowl, etc. Another application is to do a before and after sampling using disinfectant on the lab bench. The media used for these plates consists of a general-purpose medium, with some other ingredients to select for certain types of bacteria. The grid on the plate serves as an aid in counting the colonies. Below are the general guidelines for evaluating the number of colonies per plate when testing for sanitary conditions immediately after cleaning:

### Colonies per RODAC Plate

GOOD = 0-25          FAIR = 26-50          POOR = 50 and over

*Collection and Inoculation of a RODAC Sample*

1. Choose an area to which you will touch the RODAC plate. Some examples are a kitchen cutting board or counter, shower, or bathtub floor.
2. Lightly press the plate to the chosen surface.
3. Label the plate.
4. Tape it closed.
5. Incubate the plate at 30°C until the next lab period.

## PROCEDURES
Day 2 (Evaluation)

*Evaluating Your RODAC Plate Culture*

1. Examine the RODAC plate for the number of colonies.
2. If there is no fungus growing, open the plate to examine the colony types.
3. Record your results as instructed for the transport swab on the table in the evaluation portion of the Data section.

## EVALUATION OF RESULTS
## EXERCISE 13: SPECIMEN TRANSPORT

Purpose

Data

| | Surface Sampled | Total # of Colonies | # Different Colonies | Colony Description |
|---|---|---|---|---|
| Swab: | | | | |
| Colony 1 | | | | |
| Colony 2 | | | | |
| | | | | |
| | | | | |
| | | | | |
| RODAC: | | | | |
| Colony 1 | | | | |
| Colony 2 | | | | |
| | | | | |
| | | | | |
| | | | | |
| Partner's Swabs: | | | | |
| | | | | |
| | | | | |
| | | | | |
| Partner's RODAC | | | | |
| | | | | |
| | | | | |
| | | | | |
| | | | | |
| | | | | |

## CONCLUSIONS, DISCUSSIONS, AND QUESTIONS

1.  Discuss the results obtained from the different collection sites (how many types of organisms isolated, total quantity, etc).

2.  What can be concluded about microbes in the environment?

3.  In what areas of a food production plant might the RODAC plate be used?

4.  Give some examples of where specimen transport media would be used in a hospital.

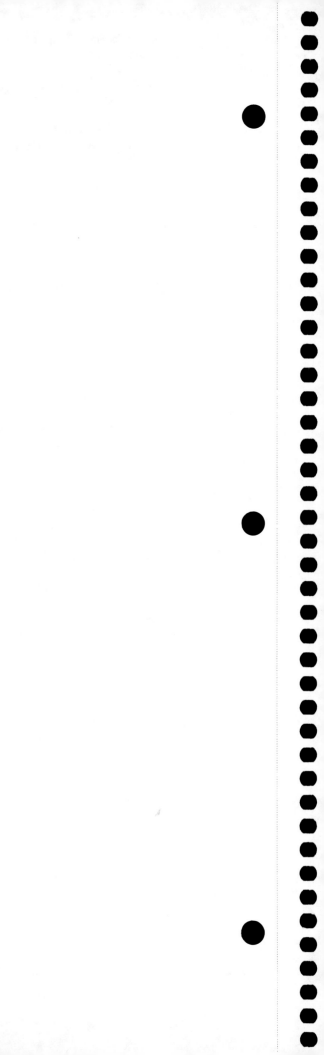

# EXERCISE *14*

## Hand-washing

### OBJECTIVES

At the conclusion of the exercise, you should...

1. understand the importance of hand-washing.
2. understand the importance of using a disinfectant when scrubbing for a hospital procedure.
3. recognize some common microorganisms found on the human skin.
4. define the terms "nosocomial," "contaminant," "transient," and "resident," when referring to microorganisms.

### INTRODUCTION

It is common for a wide variety of microorganisms to exist on the skin of the human body. Since the skin of your hands comes in contact with many different objects, it is quite natural to assume that one's hands may easily become "instruments" of potential contamination of food, wounds, etc. This becomes very important when considering the causes of nosocomial (hospital-acquired) infections. In this exercise, you will be determining the general effectiveness of hand-washing and the use of a disinfectant as a means of controlling the microflora of your hands.

### MATERIALS

Safety: Biosafety Level 1

Media:

TSA plates

Supplies:

alcohol swab

## PROCEDURES
Day 1

*Technical Background*

The microbes found on human hands may be either **transients** (contaminants) or **residents** (part of the **normal flora**). **Nosocomial** infections are infections that are spread in a hospital environment.

### *Medical Asepsis: Hand-washing*
Before hand-washing:

1. Label 3 TSA plates "Before Washing," "After Washing," and "After Alcohol."
2. Test your right hand for bacteria by firmly pressing your fingertips and fingers into the agar of the plate labeled "Before Washing."

Wash your hands with soap and water as follows:

1. Remove jewelry; push wristwatch to mid-forearm; roll or push up sleeves to mid-forearm.
2. Check hands for hangnails, cuts, or obviously dirty areas. These places need extra cleaning.
3. Turn on water with the foot pedal (or have your partner turn on the faucet).
4. Apply liquid soap and rub hands. Interface fingers to wash between them. Rub palms and backs of hands in circular motion. Pay special attention to knuckles and fingernails.
5. Continue soaping for 15 seconds.
6. Rinse with hands in down position: forearm to wrists to fingers.
7. Blot dry with towel: hands to wrists to forearms.
8. Turn off water faucet with clean, dry towel if foot pedal is not in use.
9. Test your right hand for bacteria again by firmly pressing your fingertips and fingers into the agar of the plate labeled "After Washing."
10. Disinfect the fingers of the same hand with alcohol swabs.
11. Test your hand for bacteria a third time by firmly pressing your fingertips and fingers into the agar labeled "After Alcohol."
12. Incubate all TSA plates at 35ºC until the next lab period.

*Notes:*

1. Bar soap retains bacteria; liquid soap is far more sanitary.
2. Soap is essential to remove oils, grease, and bacteria trapped on the skin.
3. Do not touch anything after washing and drying the hands. Go directly to patient care!

## PROCEDURES
Day 2

*Technical Background*

Guidelines for estimating growth:

4+ = abundant colonies on much of the area of the plate that had contact with the hand and fingers

3+ = a lot of colonies on about half of the contact area

2+ = some colonies on about 1/4 of the contact area

1+ = very few colonies

0 = no colonies

*Recording Your Results*

1. Quantify (estimate) the growth on all the plates using the guidelines listed above.
2. Describe the number and color of any isolated colonies.
3. Record all of the results in the evaluation portion of the Data section.
4. Record the results on the class boards and discuss.

# EVALUATION OF RESULTS
# EXERCISE 14: HAND-WASHING

Purpose

Data

| TSA Plate | Quantitative 1 - 4+ | Colony Description |
|---|---|---|
| Unwashed<br><br>  colony 1<br><br>  colony 2 | | |
| Partner's colony 1<br><br>Partner's colony 2 | | |
| Washed<br><br>  colony 1<br><br>  colony 2<br><br>  Partner's colony 1<br><br>  Partner's colony 2 | | |
| Disinfected<br>  colony 1<br>  colony 2<br>  Partner's colony 1<br><br>  Partner's colony 2 | | |

## CONCLUSIONS, DISCUSSIONS, AND QUESTIONS

1. From your results, how effective was hand-washing and the use of alcohol in reducing the microflora of your hand?

2. Were your results what you expected? Discuss the reasons for the results you observed.

3. Briefly define and differentiate between resident and transient microorganisms.

4. List four genera of microorganisms that can be part of the "normal flora" of the skin.

5. Name three sources of pathogenic bacteria that might contaminate someone's hands and then get transferred to patient by contact.

## Bacterial Plate Counts

### OBJECTIVES

At the conclusion of the exercise, you should...

1. be able to determine the number of bacteria in an unknown sample.
2. learn how to perform serial dilutions using serological pipettes.
3. learn how to make serial dilutions using micropipettors.
4. learn how to inoculate plates using the spread plate method.
5. be able to calculate how many bacteria are in an undiluted sample.
6. become proficient with doing dilution problems.

### INTRODUCTION

Many microbiology studies require a known number of microorganisms in a given volume. In this exercise, you will learn the standard plate count technique used to determine the number of bacteria in an unknown sample. This technique involves making dilutions of the original sample and plating the dilutions onto Petri plates using a spread plate to grow and count the bacteria present.

### MATERIALS

Safety: Biosafety Level 1

Cultures:

1 3-mL TSB culture tube of bacteria with an unknown concentration

Media:

TSA plates
99-mL water bottle

Supplies:

Pipetting device
Sterile pipettes - 1 mL
Sterile pipette - 10 mL

Colony counter
Hand tally counter
Sterile standard test tubes
Petri plate spreader
Alcohol jar for flaming
Turntable
20-200 (P200) microliter pipettor
200-1000 (P1000) microliter pipettor
Box sterile yellow pipette tips
Box sterile blue pipette tips
microcentrifuge tubes (in beaker)

## PROCEDURES
### Day 1 (Dilutions)

*Technical Background*

The **Standard Plate Count** is typically used for counting bacteria in water, milk, and food. The technique consists of diluting the bacteria using a series of test tubes containing sterile water and then plating them. It is the most widely used method to determine the number of viable cells or colony-forming units (cfu) in a culture, water, or food product sample (see diagram).

Pipettes are glass tubes made specially to aspirate (pull up) specific amounts of liquid. There are special mechanical devises used on the end of the pipette to pull the liquid in. So **NEVER PIPETTE BY MOUTH**. The accuracy of the volume measured with the pipette will determine the correct number of bacteria in a given sample.

Two types of pipetting will be introduced in this exercise. Follow the diagrams and observe the pipetting and plating demonstrations by your instructor for this procedure.

*Dilutions Using Serological Pipettes*

1. Set up 7 sterile, standard-sized test tubes in a test tube rack.
2. Label the 7 test tubes: (A) $10^{-1}$ (B) $10^{-2}$ (C) $10^{-3}$ (D) $10^{-4}$ (E) $10^{-5}$ (F) $10^{-6}$ (G) $10^{-7}$.
3. Aseptically remove a 10-mL pipette from the canister.
   a. Use the pipetting device supplied per table. (Remember, NEVER pipette by mouth.)
   b. Holding the pipette near its top, gently push the top into the device.
   c. Never touch the lower half of the sterile pipette.
   d. Use your thumb to adjust the wheel for aspirating the liquid.
4. Add 9.0 mL of sterile water from a sterile water bottle into each of the 7 test tubes.
5. Label four sterile Petri plates (remember to label them on the bottom of the plate).
   
   $10^{-5}$ = 1:100,000
   
   $10^{-6}$ = 1:1,000,000
   
   $10^{-7}$ = 1:10,000,000
   
   $10^{-8}$ = 1:100,000,000
   
   Also label the plate with date, identity of the water sample, and your names.
6. Aseptically remove a 1-mL pipette from the canister.
   a. Use the pipetting device supplied per table.

  b. Always hold the pipette near its top to gently push the top into the device.

  c. Never touch the lower half of the sterile pipette.

  d. Use your thumb to adjust the wheel for aspirating the liquid.

7. Using the 1 mL pipette, mix the broth sample carefully by pipetting up and down inside the test tube.

8. Remove 1 mL of the broth sample, and transfer it to tube A. Follow the diagram below.

9. Remove the pipette from the pipette device and place it in the pipette jar.

10. Aseptically remove a new 1-mL pipette from the canister, and attach to the device.

11. Use the pipette to mix tube A carefully, and draw up 1 mL from test tube A and transfer to tube B. Discard the pipette into the pipette jar.

12. Aseptically remove a new 1-mL pipette from the canister.

13. Use the pipette to mix tube B carefully, and draw up 1 mL from test tube B and transfer to tube C.

14. Discard the pipette as before.

15. Aseptically remove a new 1-mL pipette from the canister.

16. Use the pipette to mix tube C carefully.

17. Draw up 1 mL from test tube C and transfer to tube D.

18. Usind a new 1-mL pipette, transfer 0.1 mL (note the 0.1 markings on the pipette) from tube D $10^{-4}$ onto the plate labeled $10^{-5}$.

  a. Sterilize the Petri plate spreading rod by dipping it in the alcohol and passing it through the Bunsen burner flame. Always keep the alcohol jar well away from the Bunsen burner and replace the top after dipping the spreader.

  b. Allow the alcohol to burn off away from the flame and the alcohol jar.

  c. Hold the rod a few seconds, until it is cool.

  d. Spread the inoculum on the plate with the sterile bent glass rod, turning the plate several times for even distribution. (You may use a plate turntable, if desired.)

19. Using the same pipette as for the spread plat transfer, continue making 1:10 dilutions, as instructed, to tube G.

20. Using new pipettes for each tube, repeat the spread plate procedure for tube E $10^{-5}$ by transferring 0.1 mL to the $10^{-6}$ plate.

21. Repeat the procedure for tube F $10^{-6}$ by transferring 0.1 mL to the $10^{-7}$ plate.

22. Repeat the procedure for tube G $10^{-7}$ by transferring 0.1 mL to the $10^{-8}$ plate.

23. Invert the plates and incubate at 35°C until the next lab.

## Serological Pipetting Diagram

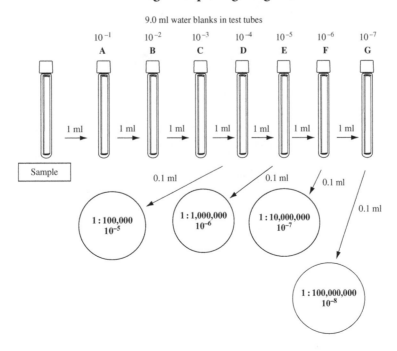

*Dilutions Using Micropipettors and Microcentrifuge Tubes*

1. Set up 7 microcentrifuge tubes in a micro tube rack (labeled A-G): (A) $10^{-1}$ (B) $10^{-2}$ (C) $10^{-3}$ (D) $10^{-4}$ (E) $10^{-5}$ (F) $10^{-6}$ (G) $10^{-7}$.
2. Using the P1000 micropipettor, and a blue tip, pipette 900 μl of sterile water into each tube.
3. Using the P200 micropipettor, and a yellow tip, pipette 100 μl of the broth sample into the first tube A.
4. Mix tube A with a new pipette tip, and remove 100 μl from tube A and place into tube B.
5. Mix tube B with a new pipette tip, and remove 100 μl from tube B and place into tube C.
6. Continue making 1/10 dilutions in this manner to tube G.
7. Label the four sterile Petri plates (remember to label on the bottom plate) $10^{-5}$ = 1:100,000; $10^{-6}$ = 1:1,000,000; $10^{-7}$ = 1:10,000,000; $10^{-8}$ = 1:100,000,000. Also label with the date, the identity of the water sample, and your names.
8. Follow the diagram below and pipette the amounts shown onto each of the labeled plates, one at a time, as follows: Using the P200 micropipettor and yellow tip, transfer 100 μl (0.1) mL from tube D onto the plate labeled $10^{-5}$. Then, dip the bent glass rod in the alcohol and flame. Hold the rod a few seconds until it is cool. Spread the inoculum on the plate with the Petri plate spreading rod, turning the plate several times for even distribution. (You may use a plate turntable, if desired.)
9. Change tips and, using the P200 pipettor, transfer 100 μl (0.1 mL) from tube E to the $10^{-6}$ plate.
10. Continue plating in this manner until the dilutions are plated.
11. Invert the plates and incubate at 35°C until the next lab.

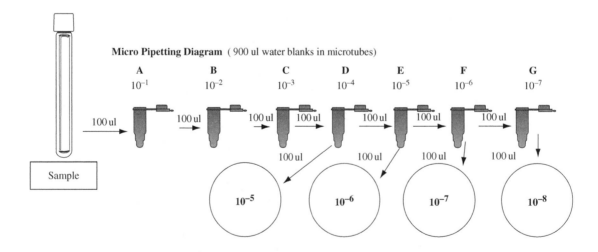

**Micro Pipetting Diagram** ( 900 ul water blanks in microtubes)

*Using Petri Plate Spreader, Alcohol Jar, and Turntable*

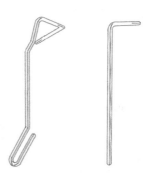

Types of Petri plate spreaders
(metal and bent glass rod).

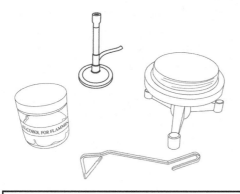

Bunsen burner, alcohol jar, turntable,
metal Petri plate spreader set up.

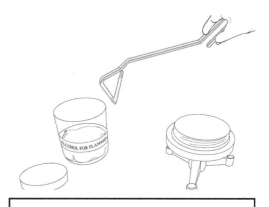

Dip Petri plate spreader into alcohol jar.

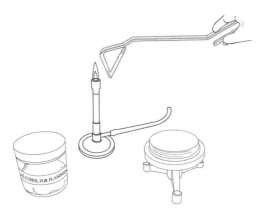

Pass Petri plate spreader through flame.
Hold away from flame for a few seconds.

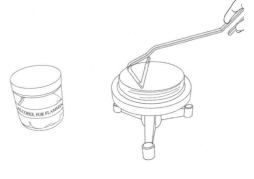

Place Petri plate on turntable, and use
spreader to spread material across plate,
while turning the turntable. Do this for at
least four to six complete turns, while
moving the spreader back and forth and
keeping it level on the agar.

# PROCEDURES
## Day 2 (Counting)

### *Technical Background*

Valid counts: Select plates that have counts that fall between 30 and 300 colonies. Counts over 300 colonies are considered invalid because of overcrowding that may cause two or more bacteria to form a single colony. Also, they are tedious to count accurately. Over 300 colonies can be recorded as "too numerous to count" (TNTC). Counts under 30 colonies (TFTC – "too few to count") are invalid because there may be a sampling error.

### *Counting Your Colonies*

1. Arrange the dilution plates in order and compare the different dilutions.
2. Select plates that have no fewer than 30 colonies and no more than 300 colonies.
3. Draw a grid on the bottom plate in quarters to help divide the plate for counting.
4. Use a colony counter or a hand counter.
5. Calculate the number of bacteria per mL in the original undiluted culture.
6. Show calculations and results in the Evaluation of Results section.
7. Use the Appendix to review doing dilutions and dilution problems.
8. Answer questions and do the dilution problems in the Evaluation of Results section.

## EVALUATION OF RESULTS
## EXERCISE 15: BACTERIAL PLATE COUNTS

Purpose

Data
**Your table's data** (every one at the table must record the data and show calculations):

| Sample | Amount Plated | Dilution | # of colonies | CFU/mL |
|--------|---------------|----------|---------------|--------|
|        |               |          |               |        |
|        |               |          |               |        |
|        |               |          |               |        |
|        |               |          |               |        |
|        |               |          |               |        |

Show calculations:

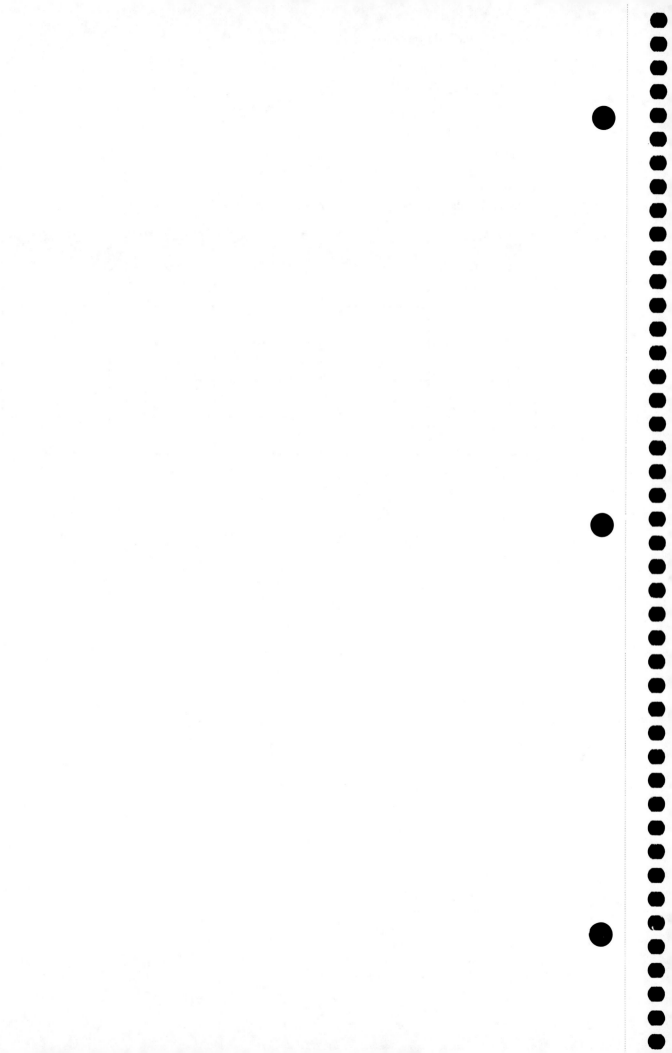

## CONCLUSIONS, DISCUSSIONS, AND QUESTIONS

1. Define the following terms:

   CFU–

   Aliquot–

   Diluent–

   Dilution factor–

   TNTC–

   TFTC–

2. Discuss the reasons why plate counts are only valid between 30 and 300 colonies.

# EXERCISE 16

## Bacterial Growth Characteristics

### OBJECTIVES

At the conclusion of the exercise, you should...

1.  understand how osmotic pressure affects a bacterial cell.
2.  learn how to determine the role of temperature and pH on bacterial growth.
3.  be able to demonstrate the oxygen requirements of bacteria.
4.  learn how to grow anaerobic bacteria.
5.  observe pigment production in bacteria.
6.  learn how to determine bacterial growth with a spectrophotometer (optional).

### INTRODUCTION

Bacteria have nutritional, physical, chemical, and environmental growth requirements. In this exercise, you will be introduced to four environmental influences on bacterial growth: osmotic pressure, temperature, pH, and oxygen.

### A. OSMOTIC PRESSURE

### MATERIALS

Safety: Biosafety Level 1

Cultures:

*Escherichia coli* - BSL1
*Staphylococcus epidermidis* - BSL1
*Halobacterium salinarium* - BSL1
*Saccharomyces cerevisiae* (yeast) - BSL1

Media:

TSB broth tube with 0.5%, 5%, 10%, 15%, and 25% NaCl

Supplies:

Turbidity measurement card set
(Optional: Spectrophotometer (Spec 20) for determining optical density measurements; see Appendix O)


## PROCEDURES
### Day 1 (Inoculations)

*Technical Background*

**Osmosis** is the process by which water crosses a membrane in response to the concentration of a **solute** (substances dissolved in water). Water will move across a cytoplasmic membrane, from a solution of low solute to a higher solute concentration, until the concentrations are equalized on both sides of the membrane. **Osmotic pressure** is the force that is exerted to maintain the concentration differences between solutions on opposite sides of the membrane. Osmotic pressure can become strong enough to cause a bacterial cell to rupture (**collapse**). For example, when Sodium chloride (NaCl) (salt) is added to a solution surrounding a bacterial cell, the concentration of solute becomes greater (**hypertonic**) outside the cell membrane and causes osmotic pressure to force water from inside the cell to the outside, thus causing the bacterial cell to collapse. Some bacteria have become adapted to environments that are hypertonic, and are able to tolerate high salt concentrations. These bacteria are called **halophiles** (salt-loving).

*Preparing Innoculations*

1. Obtain one set of the TSB with NaCl (0.5%, 5%, 10%, 15%, and 25%) tubes per table.
2. Inoculate the assigned organism into all 5 TSB broth tubes with the different concentrations of NaCl.


## PROCEDURES
### Day 2 (Results)

*Technical Background*

The following criteria will be used to evaluate the amount of growth in the TSB tubes with NaCl:

0   =   no growth (broth is completely clear)
1+   =   light growth (can see through it)
2+   =   moderate growth (can read print through it)
3+   =   heavy growth (cannot read print through it)

1. Examine all the tubes after a 48-hr. incubation.
2. Compare the inoculated tubes with the uninoculated tubes with the same concentration of NaCl.
3. Use the turbidity measurement cards to help determine the amount of growth.
4. Record all results in the table in the Evaluation of Results section.

## B. OXYGEN (CULTIVATION OF ANAEROBES)

## MATERIALS

Safety: Biosafety Level 1

Cultures:

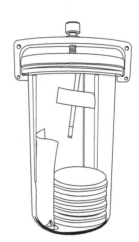

*E. coli* (facultative anaerobe) - BSL1
*Bacillus subtilis* (strict aerobe) - BSL1
*Clostridium sporogenes* (aerotolerant anaerobe) - BSL1
*Clostridium species* (an obligate anaerobe) - BSL1
*Microaerophilic anaerobe* (if available) - BSL1

Media:

Brain Heart Infusion plates
Thioglycollate broths
Anaerobe jar
Gas pack set up
Methylene blue indicator strip

## PROCEDURES
Day 1 (Inoculations)

*Technical Background*

Oxygen is extremely important for microbial growth. Some microorganisms cannot live without it. These microbes are called **strict aerobes**. Some microbes cannot live if oxygen is present. These are called **strict (obligate) anaerobes**. Oxygen is toxic to obligate anaerobes. In between the two extremes are the **facultative anaerobes**. They can grow with or without oxygen, but grow better with oxygen. **Aerotolerant anaerobes** cannot use oxygen to grow, but can tolerate it. Their growth will be enhanced by microaerophilic conditions. **Microaerophilic** bacteria grow best when the atmosphere has increased $CO_2$ (carbon dioxide) and lower concentrations of oxygen are present.

There are special techniques for growing anaerobes. One is to use a reducing medium that contains chemicals that combine with free oxygen and decrease the concentration of oxygen.

**Thioglycollate** media has the reducing chemical sodium thioglycollate and a small amount of agar added, which reduces the diffusion of oxygen into the medium. Sometimes an indicator is used in the thioglycollate medium to indicate if oxygen is present. The reducing agent **resazurin** is pink in the presence of oxygen and colorless when reduced.

Conventional media can be used to grow anaerobes when incubated in an anaerobic environment. An anaerobic environment can be made with a **Brewer anaerobic jar**. The anaerobic jar is used to remove oxygen from a sealed container by catalyzing the chemical combination of oxygen with hydrogen to form water. The anaerobic jar consists of using either a gas pack that has sodium bicarbonate and sodium borohydride, along with the catalyst palladium or an AnaeroGen$^{TM}$ sachet. A **methylene blue indicator strip** is added to detect the absence of oxygen. The chemical methylene blue is blue in the presence of oxygen and white in the absence of oxygen.

**Gas Pack:** Water is added to the gas pack, and the reaction that takes place in the anaerobic jar is:

$$2\,H_2 + O_2 \quad \text{catalyst} \quad 2\,H_2O$$

**AnaeroGen$^{TN}$ by Oxoid:** The sachet is placed in the anaerobe jar along with the methylene blue indicator strip, and the atmospheric oxygen is absorbed with the simultaneous generation of carbon dioxide. It is a novel method that differs from the gas pack method in that the reaction does not generate any hydrogen and, therefore, does not need a catalyst or require the addition of water to the pack.

After about 2-3 hours, the jar will become anaerobic, the methylene blue strip should be completely white, and there should be moisture droplets on the inside of the jar. Another gas pack that is available does not need water and will react with the air to produce an anaerobic environment. Other methods for obtaining anaerobic culture conditions include anaerobic incubators and glove boxes. The air is evacuated from the chamber and replaced with a mixture of carbon dioxide and nitrogen.

*Growing Aerobic and Anaerobic Bacteria*

1.  Label the thioglycollate tubes with the names of the bacteria used.
2.  Inoculate one loopful of *E. coli* into a thioglycollate tube.
3.  Repeat the inoculation of all of the bacteria listed into the correctly labeled thioglycollate tubes.
4.  Incubate the thioglycollate test tubes (loose caps) at 35°C until the next lab period.
5.  Divide the two Brain Heart infusion agar plates into fourths. (See diagram.)
6.  Label one plate "Air" and the other "Anaerobic."
7.  Label the quadrants with the four species of bacteria.
8.  Inoculate each quadrant with the appropriate bacterium. Use one streak. (See the diagram.)
9.  Invert and incubate the air plate at 35°C until the next lab period.
10. Invert the anaerobic plate and place it in the anaerobic jar.
11. After all plates are in the anaerobic jar, the instructor will demonstrate the technique for setting up the anaerobic jar.
12. Incubate the anaerobic jar at 35°C until the next lab period.

Diagram for inoculating the
BHI plates (4 organisms)

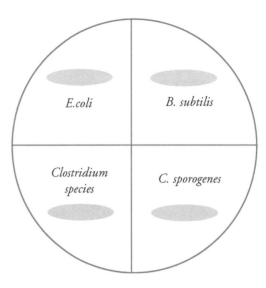

Diagram for inoculating BHI plate
(3 organisms)

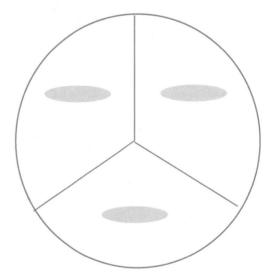

## PROCEDURES
Day 2 (Results)

*Technical Background*

**Thioglycollate interpretations**: Strict aerobes will grow mostly on the surface of the medium where there is more oxygen. Obligate anaerobes will grow mostly at the bottom where there is no oxygen. Facultative anaerobes will grow throughout the medium. Microaerophiles will grow below the surface where there is less oxygen, but not in the bottom, where there is no oxygen. Aerotolerant anaerobes will grow heavier at the bottom, but will also grow throughout the media. The incubation time can affect the results. The faster-growing facultative anaerobic bacteria will grow rapidly, die, and fall to the bottom of the tube. This should not be interpreted as strict anaerobic growth. There will still be increased growth throughout the thioglycollate medium. See the diagram for interpreting thioglycollate media.

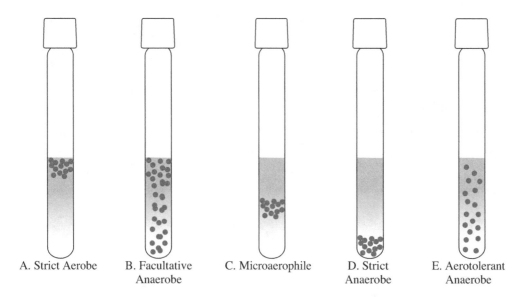

A. Strict Aerobe    B. Facultative        C. Microaerophile    D. Strict          E. Aerotolerant
                      Anaerobe                                     Anaerobe           Anaerobe

**Brain Heart Infusion plate interpretations:** Strict aerobes should grow only on the air plate. Obligate anaerobes should grow only on the anaerobic plate. Facultative anaerobes should grow on both the air and anaerobic plates. Aerotolerant anaerobes should have more growth on the anaerobic plate.

*Evaluating Your Bacteria Growth*

1. Using the descriptions and diagrams, read and record the results of the thioglycollate tubes in the data section of the Evaluation of Results.
2. Obtain the BHI plates from the anaerobic jar and air incubator.
3. Read and record the results in the data section of the Evaluation of Results. Compare your results to descriptions of the oxygen requirements for each organism.

# C. PH

# MATERIALS

Safety: Biosafety Level 1

Cultures:

*E. coli* - BSL1
*Alcaligenes faecalis* - BSL1
*S. epidermidis* - BSL1
*S. cerevisiae* - BSL1

Media:

TSB test tube, pH 5
TSB test tube, pH 7
TSB test tube, pH 9

Supplies:

1 rack of uninoculated tubes each with a different, but standard, turbidity
Turbidity measurement card set
(Optional: Spectrophotometer (Spec 20) for determining optical density measurements; see Appendix L)

## PROCEDURES
### Day 1 (Inoculations)

*Technical Background*

The acidity or alkalinity (**pH**) of the environment influences microbial growth. There are some microbes that can grow at acidic pH values below 4.0 or at alkaline pH values above 8.0. Most bacteria grow best between pH 6.5 and 7.5, and are inhibited below about 4.5 or above 8.5. Microbial growth can be measured as colonies on a plate or the degree of turbidity in a broth culture. Turbidity does not indicate the number of microbes directly, but increasing turbidity indicates increasing numbers. In this exercise, microbes will be inoculated into trypticase soy broth of different pH values, and the amount of growth estimated after 48 hours of incubation.

*Preparing pH Solutions for Bacteria Growth*

1. Each pair of students will be assigned an organism.
2. Use aseptic technique to inoculate the assigned organism into the pH 5 TSB tube.
3. Repeat with the pH 7 and pH 9 TSB tubes.
4. Incubate all TSB tubes (loose caps) at 35°C until the next lab period.

## PROCEDURES
### Day 2 (Results)

*Evaluating the Effect of pH Solutions*

1. Tighten the caps and vortex each tube.
2. Do a visual reading of each broth by comparing each to the standard tubes and using the measurement cards:
   - 0 = no growth (broth is completely clear)
   - 1+ = light growth (can see through it)
   - 2+ = moderate growth (can read print through it)
   - 3+ = heavy growth (cannot read print through it)
3. Record your results and those of other students in the data table in the Evaluation of Results section.

## D. TEMPERATURE

## MATERIALS

Safety: Biosafety Level 1

Cultures:

*Escherichia coli* - BSL1
*Geobacillus stearothermophilus* (old name = *Bacillus stearothermophilus*) - BSL1
*Psychrobacter urativorans* (old name = *Micrococcus cryophilus*) - BSL1
*Saccharomyces cerevisiae* - BSL1

Media:

5 TSB test tubes (for 5 temperatures)

Supplies:

Incubators set at 35°C, 42°C, 55°C
Refrigerator for 4-5°C
In room for 25°C (room temperature)

## PROCEDURES
Day 1 (Inoculations)

*Technical Background*

Temperature is one of the most important environmental influences on microbial growth. Most microbes grow within a specific temperature range. The **minimum growth temperature** is the lowest temperature at which a microbial species will grow, and the **maximum growth temperature** is the highest at which a species will grow. The **optimum growth temperature** is the temperature at which a species will grow the fastest. **Psychrophilic** bacteria grow between 0°C and 5°C. **Mesophilic** bacteria grow best between 25°C and 40°C. **Thermophiles** can grow at temperatures between 45°C and 65°C.

Most bacteria are adversely affected by temperatures above 50°C, but the length of time required to kill them varies. Therefore, the time required to kill bacterial cells is essential to know for safety reasons.

*Preparing Bacteria Innoculations at Varying Temperatures*

Each pair of students will be assigned the organism to use for the inoculations.
   1.   Using aseptic technique, inoculate a loopful of the assigned organism into each of the 5 TSB test tubes.

2. Incubate the inoculated TSB tubes into each of 5 different temperatures until the next lab period.
3. Place all of the inoculated tubes into a test tube rack labeled for your section and incubate as follows:
   4-5°C = refrigerator
   25°C = room temperature (in classroom on shelves)
   37°C = section incubator labeled 37°C
   42°C = shared incubator labeled 42°C
   55°C = shared incubator labeled 55°C

## PROCEDURES

## Day 2 (Results)

*Evaluating the Effect of Temperature on Bacteria Growth*

1. Tighten the caps and vortex each tube.
2. Do a visual reading of each broth by comparing each to the standard tubes and using the measurement cards as above. Use the following guidelines for the visual readings (as above):
   0 = No growth (broth is completely clear)
   1+ = Light growth (can see through it)
   2+ = Moderate growth (can read print through it)
   3+ = Heavy growth (cannot read print through it)
3. Record the results in the Evaluation of Results section.

## E. PIGMENT PRODUCTION

## MATERIALS

Safety: Biosafety Level 1

Cultures:

*Staphylococcus epidermidis* - BSL1
*Pseudomonas aureofaciens* - BSL1
*Serratia marcescens* - BSL1
*E. coli* (control) - BSL1

Media:

TSA plates (1 TSA plate per pair)

Supplies:

30°C incubator
35°C incubator

## PROCEDURES
### Day 1 (Inoculations)

*Technical Background*

Some species of bacteria are capable of pigment production. Pigment production is included when studying growth characteristics of bacteria and is often used as a clue for identifying unknown bacteria. Isolated colonies are the best way to observe whether the pigment is soluble or nonsoluble. A soluble pigment is water soluble, and will diffuse into the surrounding medium. The medium will turn the color of the pigment that is produced. *Pseudomonas aeruginosa* produces a green pigment that will diffuse into the media. Nonsoluble pigments remain inside the bacterial cells and only the colony is colored. *Pseudomonas aureofaciens* produces a non-soluble yellow to orange pigment. *Micrococcus luteus* colonies are bright yellow. Pigment production is controlled by enzyme activity and can be affected by different temperatures. For example, *Serratia marcescens* produces a red pigment at 30°C, but not at 35°C.

*Using Pigment Production to Identify Bacteria*

1. Divide the TSA plates into fifths.
2. Label one TSA plate 30°C.
3. Label the other TSA plate 35°C.
4. Inoculate the quadrants on each plate with one each of the bacteria provided.
5. Streak each organism for isolation. (See diagram.)

6. Invert and incubate one plate at 30°C and the other at 35°C until the next lab period.

## PROCEDURES
Day 2 (Results)

*Evaluating Pigment Production*

1.  Examine both plates for pigment production.
2.  Note whether the pigment is different at the two temperatures.
3.  Note whether the pigment has diffused into the agar.
4.  Record the results in the Evaluation of Results section.

## EVALUATION OF RESULTS
## EXERCISE 16: BACTERIAL GROWTH CHARACTERISTICS

Purpose

## PART A: OSMOTIC PRESSURE RESULTS

| Culture: | 0.5% NaCl | 5% NaCl | 10% NaCl | 15% NaCl | 25% NaCl |
|---|---|---|---|---|---|
| E. coli | | | | | |
| S. epidermidis | | | | | |
| S. cerevisiae | | | | | |
| Halobacterium | | | | | |

Data

## CONCLUSIONS, DISCUSSIONS, AND QUESTIONS

1.  Discuss some reasons for the results that were obtained.

2.  What organism studied is the most tolerant of high concentrations of salt? Where in nature would they be found?

# PART B: OXYGEN RESULTS

Data

Brain heart Infusion Plates:

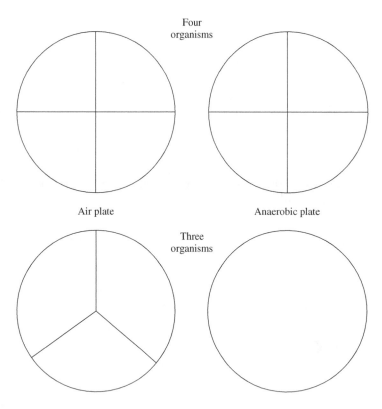

Four
organisms

Air plate                                    Anaerobic plate

Three
organisms

Thioglycollate tubes:

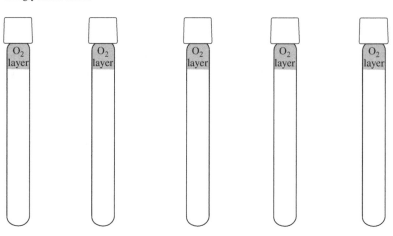

## CONCLUSIONS, DISCUSSIONS, AND QUESTIONS

1. Discuss some reasons why the results might not be textbook perfect.

2. Draw and label the parts of an anaerobic jar set up.

## PART C: pH RESULTS

Data

Record amount of visual growth by gently shaking the tube and estimating growth by comparison of its turbidity to standard tubes.

 0  = no growth
1+ = light growth (can see through it)
2+ = moderate growth (can read print through it)
3+ = heavy growth (cannot read print through it)

| Species | Amount of growth at: | | |
|---|---|---|---|
| | pH 5 | pH 7 | pH 9 |
| E. coli | | | |
| A. faecalis | | | |
| S. epidermidis | | | |
| S. cerevisiae | | | |

## CONCLUSIONS, DISCUSSIONS, AND QUESTIONS

1. Can you determine the optimum pH for each of the organisms tested?

2. Which of the four species studied grows in an acidic environment? Where would you expect these organisms to live in humans?

3. Name four different human foods that are preserved using pH values below pH 5.

## PART D: TEMPERATURE RESULTS

Data

After 48 hours of growth at each temperature, estimate the amount of visual growth by comparing the vortexed culture to the standard tubes:

     0   = no growth (broth is completely clear)
     1+ = light growth (can see through it)
     2+ = moderate growth (can read print through it)
     3+ = heavy growth (cannot read print through it)

| Amount of growth at temperature of incubation | | | | | |
|---|---|---|---|---|---|
| **Species** | **4-5° C** | **25° C** | **35°** | **42° C** | **55° C** |
| *E. coli* | | | | | |
| *Saccharomyces cerevisiae* | | | | | |
| *Geobacillus stearothermophilus* | | | | | |
| *Psychrobacter urativorans* | | | | | |

## CONCLUSIONS, DISCUSSIONS, AND QUESTIONS

1. Based on the results obtained, determine the optimal growth temperatures for the organisms studied.

2. Discuss reasons why it is helpful to know the temperatures at which bacteria grow.

3. What is the optimum temperature of human pathogens?

4. Define the following terms:

   Thermophile–

   Mesophile–

   Psychrophile–

## PART E: PIGMENT PRODUCTION

Data

| Temperature of incubation | E. coli Color of: | | S. epidermidis Color of: | | P. aureofaciens Color of : | | S. marcescens Color of: | |
|---|---|---|---|---|---|---|---|---|
| | colony | agar | colony | agar | colony | agar | colony | agar |
| 30° C | | | | | | | | |
| 35° C | | | | | | | | |

## CONCLUSIONS, DISCUSSIONS, AND QUESTIONS

1. Which temperature seems optimum for pigment formation (if any) by
   *Serratia marcescens?*
   *Pseudomonas aureofaciens?*
   *S. epidermidis?*
   *E. coli?*

2. You are taking care of a patient with severe burns. You notice while changing the bandages, that the pus is bluish-green and smells of grape Kool-Aid. What should you report to the attending physician?

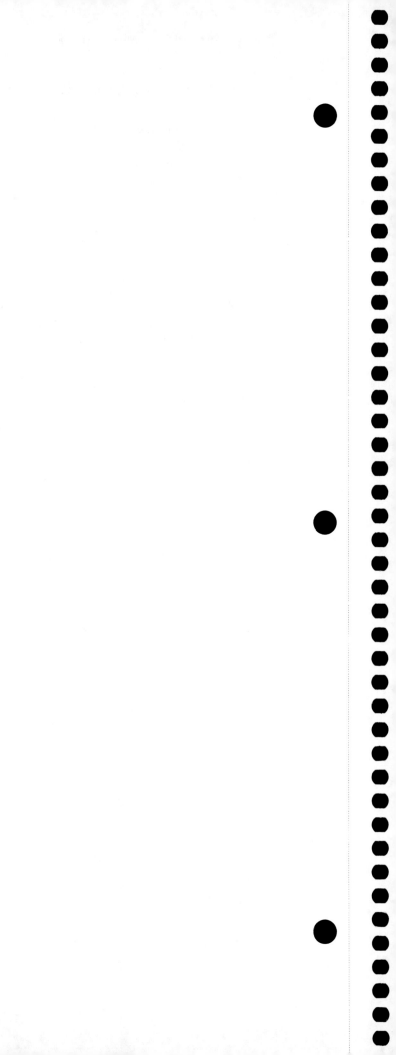

# E X E R C I S E   *17*

## Selected Physiological and Biochemical Characteristics (Introducing the Gram-Negative Rods)

### OBJECTIVES

At the conclusion of the exercise, you should...

1. be able to perform and interpret select biochemical tests on bacteria.
2. become familiar with some of the Gram-negative rods in the Enterobacteriaceae and the Pseudomonadaceae families.
3. be able to use the physiological and biochemical tests in this exercise to identify Gram-positive and Gram-negative unknown bacteria.

### INTRODUCTION

Many bacteria have the same colony and cellular morphology. In order to classify them, other tests are necessary. These tests involve the metabolism of bacteria and are based on which substrates bacteria use and what metabolic products are formed. Laboratory tests have been developed to identify the products of the metabolic pathways.

In this exercise, you will perform selected tests that will aid in classifying and identifying bacteria. The family Enterobacteriaceae encompasses the most frequently encountered single group of bacteria in the clinical laboratory. The members inhabit the gastrointestinal tract of humans and some animals. They are also found in the environment. The family consists of species of Gram-negative, oxidase-negative, straight rods. They all ferment glucose. There are many types of diseases that these organisms cause; for example, pneumonia, urinary tract infections, wound infections, meningitis, and bacteremia. *Escherichia coli* is the most well-known member of this family.

The family Pseudomonadaceae includes species from the Genera *Pseudomonas, Xanthomonas, Frateuria,* and *Zoogloea.* Of these species, Pseudomonas is the most well-known Genus and contains species that are pathogenic to man. All are Gram-negative, aerobic, catalase-positive rods. They can be separated from the enteric and other fermentative bacteria by their strictly oxidative character.

## A. CASEIN HYDROLYSIS

## MATERIALS

Cultures:

*Pseudomonas stutzeri* (negative) BSL1
*Bacillus* species (positive) BSL1

Media:

Casein (skim milk) plate

## PROCEDURES
Day 1 (Inoculations)

*Technical Background*

Casein is a protein in milk. Some bacteria produce an enzyme named caseinase, which hydrolyzes casein. When they do, a clear zone develops around the streak on a casein plate. This clear zone indicates positive casein hydrolysis.

*Preparing Inoculations*

1. Streak a loopful of Bacillus species onto one-half of a casein plate, in a single line (see diagram).
2. Repeat the same streaking procedure with the other organism provided.
3. Invert the plate and incubate at 35ºC until the next lab period.

Heavy streak of bacteria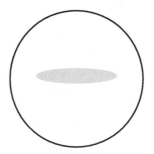

## PROCEDURES
Day 2 (Results)

*Evaluating Your Results*

1. Examine each of the casein plates for a clear zone surrounding the streaked area. (See demo.)
2. Record your results, as well as those from other students, in the table in the Evaluation of Results section.

## B. CATALASE

## MATERIALS

Safety: Biosafety Levels 1 and 2

Cultures:

*E. coli* TSA slant (positive) - BSL1
*Streptococcus* species (negative) - BSL2

Supplies:

3% Hydrogen peroxide (fresh)
Microscope slides

## PROCEDURES
Day 1

*Technical Background*

Most aerobic and facultative bacteria utilize oxygen, which produces **hydrogen peroxide**, which is toxic to the bacteria's own enzyme systems. They can survive because they can produce an enzyme called **catalase**, which can convert the hydrogen peroxide to water and oxygen.

To determine whether the enzyme catalase is produced, a few drops of hydrogen peroxide is placed on the organism. If the organism produces **bubbles** immediately, it is **catalase positive**. No bubbles, **catalase negative**.

The Gram-negative rods in the family Enterobacteriaceae (referred to as the **Enteric bacteria**) are catalase positive. *E. coli* is the one of the most well-known enteric bacterium.

This test separates two Genera that will be studied in later exercises (Staphylococci and Streptococci).

*Testing Bacteria for Catalase Reaction*

I.  Slide Method (behind shield):
    1.  Transfer a loopful of E. coli from the TSA slant onto a microscope slide.
    2.  Add 1 drop of 3% hydrogen peroxide.
    3.  Watch for the immediate release of bubbles (positive test).
    4.  No bubbles is a negative test.
    5.  Repeat the same procedure with the other bacteria provided.

II. Tube method:
    1.  Add 4 to 5 drops of 3% $H_2O_2$ to a 12 x 75-mm test tube.
    2.  Using a wooden applicator stick, collect a small amount of organism from a well-isolated 18- to 24-hour colony and place into the test tube. Be careful not to pick up any agar. This is particularly important if the colony isolate was grown on agar containing red blood cells. Carryover of red blood cells into the test may result in a false-positive reaction.
    3.  Place the tube against a dark background and observe for immediate bubble formation ($O_2$ + water = bubbles) at the end of the wooden applicator stick.
    4.  Positive reactions are evident by immediate effervescence (bubble formation) (Fig. 2A). Use a magnifying glass or microscope to observe weak positive reactions.
    5.  No bubble formation (no catalase enzyme to hydrolyze the hydrogen peroxide) represents a catalase-negative reaction.
    6.  Quality control is performed by using organisms known to be positive and negative for catalase.

Reference: http://www.microbelibrary.org/index.php/library/laboratory-test/3226-catalase-test-protocol

# C. CITRATE

## MATERIALS

Safety: Biosafety Level 1

Cultures:

*E. coli* (negative) - BSL1
*Enterobacter aerogenes* (positive) - BSL1

Media:

Simmon's citrate slants

## PROCEDURES
### Day 1 (Inoculations)

*Technical Background*

Some bacteria are capable of getting energy from citrate (citric acid) as their sole source of carbon. They use the enzyme citritase or citrate demolase. Therefore, the bacteria can utilize the citric acid, grow, and metabolize an excess of sodium and ammonium ions. This produces alkaline conditions, which turns the indicator from green to blue. The citrate test aids in the differentiation between some genera of bacteria.

*Preparing Inoculations*

1. Label the the two citrate slants with one each of the organisms provided. Inocuate by starting at the bottom of the slant and streaking back and forth, while moving up the slant.
2. Incubate at 35°C (loose caps) until the next lab period.

## PROCEDURES
### Day 2 (Results)

*Evaluating Your Results*

1. A positive citrate test is indicated by growth with an intense blue color on the slant.
2. A negative citrate test is indicated by no growth, with no change in color. The slant remains green.
3 Record the citrate results in the Evaluation of Results section.

## D. CARBOHYDRATE FERMENTATION
## MATERIALS

Safety: Biosafety Level 1

Cultures:

*E. coli* (positive) - BSL1
*Pseudomonas stutzeri* (negative) - BSL1

Media:

BCP lactose broths

## PROCEDURES
Day 1 (Inoculations)

*Technical Background*

The principle of carbohydrate fermentation tests is to determine the ability of an organism to ferment (degrade) specific carbohydrates (sugars) that are incorporated into a basal medium, producing acid or acid with visible gas, by using specific enzymes. Fermentation studies are generally characteristic for specific bacterial groups or species. For example, all members of the Enterobacteriaceae family ferment glucose. Lactose fermentation is often used to distinguish certain pathogenic enteric species (*Salmonella* and *Shigella* are lactose negative). Mannitol fermentation is used to differentiate *Staphylococcus aureus* from *Staphylococcus epidermidis*.

Purple broth is the base medium used for the fermentation tests in this manual. Brom cresol purple is the indicator that turns yellow in the presence of acids and remains purple in a neutral or alkaline environment. **Durham tubes** are sometimes used to determine if gas has been produced as an end product of metabolism. This does not tell you what kind of gas is produced. Gas production from glucose is sometimes useful in identifying some members in the family Enterobacteriaceae. Some examples of the BCP sugars that can be tested in this manner are Adonitol, P Lactose, Mannitol, Raffinose, and Sucrose.

*Preparing Inoculations*

1.  Label the BCP lactose tube with the organism to be inoculated.
2.  Inoculate one loopful of each of the organisms provided into the appropriate tube.
3.  Incubate at 35°C (loose caps) until the next lab period.

## PROCEDURES
Day 2 (Results)

*Evaluating Your Results*

1.  Examine the BCP lactose fermentation tubes for acid production. A yellow color is indicative of a positive test, purple color is negative. If Durham tubes were included, a bubble in the tube is indicative of gas production.
    **Note:** Some bacteria actually consume the indicator dye turning the broth gray or brown. In this case, have a drop of Brom cresol purple added to the test tube, mix and read color as usual; purple = negative, yellow = positive.

2.  Record all results in the Evaluation of Results section.

## E. HYDROGEN SULFIDE PRODUCTION

## MATERIALS

Safety: Biosafety Level 1

Cultures:

*E. coli* (negative) - BSL1
*Citrobacter freundii* (positive) - BSL1

Media:

PIA stabs

## PROCEDURES
Day 1 (Inoculations)

*Technical Background*

The principle of the hydrogen sulfide ($H_2S$) test is to determine whether an organism produces the enzyme cysteine desulfurase, which results in the liberation of $H_2S$ from sulfur-bearing amino acids. The $H_2S$ combines with the iron in the media, producing a visible black color. Peptone iron agar is used as the media for hydrogen sulfide production, and contains ferric salt as the indicator. When sulfide is produced, it reacts with the ferric metal salt to produce a visible black precipitate ($H_2S$). The test aids in the differentiation of certain species of enteric bacteria. Some species of *Proteus* and *Citrobacter* produce $H_2S$, whereas others do not. *E. coli* does not produce hydrogen sulfide. Also, $H_2S$ production is a key test in differentiating *Shigella* (negative) from *Salmonella* (positive).

*Preparing Inoculations*

1. Label the PIA tubes with the organisms to be inoculated.
2. Inoculate the media by obtaining a sample of the organism with the inoculating needle and stabbing it down to the bottom of the tube.
3. Incubate at 35°C (loose caps) until the next lab period.

## PROCEDURES
Day 2 (Results)

*Evaluating Your Results*

1.  Examine the PIA stabs for blackening of the agar. Any black color within the media is considered positive; no change in color is negative.
2.  Record the results in the Evaluation of Results section.

## F. INDOLE PRODUCTION

## MATERIALS

Safety: Biosafety Level 1

Cultures:

*E. coli* (positive) - BSL1
*Enterobacter aerogenes* (negative) - BSL1

Media:

Tryptone broth tubes

Supplies:

Kovac's reagent (fresh)

## PROCEDURES
Day 1 (Inoculations)

*Technical Background*

The principle of the indole test is to determine the ability of an organism to split indole from tryptophan. Tryptophan is an amino acid that can be oxidized by certain bacteria to form indole. The enzyme tryptophanase is produced by certain species of bacteria, which deaminates the tryptophan molecule forming indole, pyruvic acid, and ammonia. After 24-48 hr. of growth, a reagent is added to the tryptone broth to detect the formation of indole. If indole has been produced, then a red (or a bright pink or magenta color) ring will form on top of the tryptone broth. If no indole has been produced, then a yellow ring will form on top of the tryptone broth. This test is useful in the differentiation of *E. coli* from other enteric bacterial species. It also aids in differentiating *Proteus mirabilis* from other *Proteus* species.

*Preparing Inoculations*

1. Label and inoculate the tryptone broths with a loopful of each of the bacteria provided.
2. Incubate at 35°C (loose caps) until the next lab period.

## PROCEDURES
Day 2 (Results)

*Technical Background*

Kovac's is a chemical reagent that is added to the tryptone broth after growth has occurred. A red layer should form on top of the culture for a positive test result. A yellow layer indicates a negative result. If the test is negative, the original broth tube can be reincubated and retested.

*Testing with Kovac's Reagent*

1. After growth has occurred at 24-48 hours incubation, remove 1 mL of tryptone broth aseptically, with a 1-mL pipette.
2. Transfer the 1 mL to a serology tube.
3. Add 3-5 drops of Kovac's reagent to the tryptone broth tube.
4. Gently shake the serology tube, while keeping the Kovac's reagent on the surface of the broth.
5. Examine the tube for a red (positive) or yellow (negative) ring at the surface of the broth.
6. Record the results in the Evaluation of Results section.
7. If a positive test result is obtained, then discard both test tubes.
8. If a negative test is obtained, then reincubate the original tryptone broth tube for another 24-48 hr. and retest as above.
9. Discard the serology tube with the reagents.
10. When the test is completed and confirmed negative or positive, discard all contaminated test tubes properly.

## G. METHYL RED TEST

## MATERIALS

Safety: Biosafety Level 1

Cultures:

*E. coli* (positive) - BSL1
*Enterobacter aerogenes* (negative) - BSL1

Media:

Methyl Red broth tubes

Supplies:

Methyl red indicator

## PROCEDURES
Day 1 (Inoculations)

*Technical Background*

The methyl red test is used to test the ability of certain bacteria to produce large amounts of acid from glucose fermentation (mixed acid fermentation), utilizing the enzyme formic hydrogenylase. The MR broths contain glucose and if an organism produces a large amount of organic acid from glucose (pH below 5), then the medium will turn red when the methyl red reagent is added. If the by-products are neutral, then the methyl red will turn yellow, indicating a pH above 6.0. The test can be useful in differentiating *Klebsiella* species and *Enterobacter aerogenes* (both usually negative) from *E. coli* (usually positive).

*Preparing Inoculations*

1.  Label and inoculate the methyl red broth tubes with a loopful of each of the bacteria available.
2.  Incubate at 35°C (loose caps) until the next lab period.

## PROCEDURES
Day 2 (Results)

*Evaluating Your Results*

1.   After growth has occurred at 24-48 hours incubation, remove 1 mL of the MR broth aseptically with a 1-mL pipette.
2.   Transfer the 1 mL to a serology tube.
3.   Add 3-5 drops of the methyl red indicator reagent to the culture.
4.   Gently shake the serology tube, mixing the broth and reagent.
5.   If the indicator turns red, then the organism is a mixed acid fermenter and is positive for the methyl red test.
6.   A yellow color indicates that neutral products were produced, and the test is considered negative.
7.   Record the results in the Evaluation of Results section.
8.   If a negative test is obtained, then reincubate the original MR broth tube for another 24-48 hr. and retest as above.
9.   Discard the serology tube with the reagents.
10.  When the test is completed and confirmed negative or positive, discard all contaminated test tubes correctly.

# H. MOTILITY

## MATERIALS

Safety: Biosafety Level 1

Cultures:

*E. coli* (positive) - BSL1
*Rahnella aquatilis* (negative) - BSL1
*S. epidermidis* (negative) - BSL1

Media:

Motility semi-solid agar tubes (stabs)

## PROCEDURES
Day 1 (Inoculations)

*Technical Background*

Bacteria can be motile or nonmotile. Some bacteria are motile by means of flagella. Whether bacteria are motile or nonmotile can be useful in classifying them. There are different techniques to distinguish motility. Some are by direct observation of the bacteria with a microscope slide wet mount preparation. The inexperienced observer must learn to tell the difference between motility and Brownian movement. Another technique is to inoculate the bacteria into a semi-solid (soft) agar tube. An inoculating needle is used to stab into the soft agar. After incubation, the inoculated agar tube is compared to an uninoculated tube. If there is growth away from the stab area, then the organism is considered to be motile (positive motility test). This technique can be difficult to interpret also, but becomes easier with experience.

*Preparing Inoculations*

1. Label the motility media tubes with the bacteria to be inoculated.
2. Make sure the inoculating needle is straight.
3. Flame and cool the inoculating needle, and obtain a sample of the organisms on the tip.
4. Inoculate the bacteria into the media, by stabbing the media to the bottom of the tube. Be sure to stab the media as straight in and out as possible, without moving the needle around.
5. Incubate the motility tubes at 35°C (loose caps) until the next lab period.

## PROCEDURES
Day 2 (Results)

*Evaluating Your Results*

1.  Motility is determined by looking for growth away from the stab line.
2.  Hold the inoculated tube up to the light and compare it to an uninoculated tube.
3.  Cloudiness away from the stab indicates motility.
4.  Growth at the top should not be interpreted as positive.
5.  If it is too difficult to interpret, aseptically transfer a loopful of the bacteria from the growth at the top to a microscope slide and make a wet mount.
6.  Observe the wet mount for actively swimming bacteria.
7.  Record the results the Evaluation of Results section.


# I. NITRATE REDUCTION

## MATERIALS

Safety: Biosafety Level 1

Cultures:

*E. coli* (positive)
*Alcaligenes faecalis* (negative)

Media:

Nitrate broth tubes

Supplies:

Nitrate reagents A & B

## PROCEDURES
### Day 1 (Inoculations)

*Technical Background*

The nitrate reduction test is used to determine the ability of an organism to reduce nitrate to nitrites or free nitrogen gas. It involves the enzyme nitratase. The reduction of nitrate ($NO_3^-$) to nitrite ($NO_2^-$) and to nitrogen gas ($N_2$) takes place under anaerobic conditions, where an organism utilizes nitrate as its final electron acceptor instead of oxygen. The reduction of nitrate can be detected by testing for the presence of nitrite after bacteria have grown in the broth. Depending on which enzymes the bacterium is capable of producing, one of three reactions can occur: 1. nitrate is reduced to nitrite; 2. nitrate is rapidly reduced to nitrite, then further reduced to ammonia or free nitrogen; or 3. The nitrate remains unaltered. To test the first reaction above, reagents are added, and if nitrate is reduced to nitrite, then a red color develops. This is a positive reaction. No further testing is necessary. If the test is negative (no color change), then an additional test needs to be done to distinguish between the two remaining possibilities (if the nitrate remained unaltered (negative), or has been reduced to ammonia or free nitrogen (positive)). This is accomplished by adding zinc dust, which will chemically reduce nitrate to nitrite. If the nitrate remains unaltered by the bacterial growth, then the zinc will reduce this nitrate to nitrite and a red color will develop (negative nitrate result). If no color change occurs after adding the zinc dust, the nitrate has been completely reduced by the bacteria beyond the nitrite stage to $NH_3$ or $N_2$, indicating a positive result. Another indication that $N_2$ gas has been produced is a significant amount of gas production in the Durham tube. **Durham tubes** are small tubes placed upside down in the test tube. The purpose of the inverted tube is to show if gas has been produced as an end product of metabolism. This does not tell you what kind of gas is produced.

Nitrate Reduction and Denitrification are carried out by bacteria capable of anaerobic respiration, using the nitrate as terminal electron acceptors. When nitrite is the end product, it accumulates in the growth medium and eventually inhibits growth, as it is fairly toxic. Denitrifiers are able to further reduce the nitrite to nitrogen gas and thereby relieve the nitrite inhibition of growth. All bacteria that are able to use nitrate for anaerobic respiration prefer to use oxygen instead, and will only express the genes for nitrate reduction/denitrification in the absence of oxygen.

*Preparing Inoculations*

1. Label the nitrate tubes appropriately.
2. Inoculate a loopful of each of the bacteria into the nitrate tubes.
3. Incubate at 35°C (loose caps) until the next lab period.

## PROCEDURES
Day 2 (Results)

*Evaluating Your Results*

1.  After growth has occurred, remove 1 mL of the nitrate broth aseptically, with a 1-mL pipette.
2.  Transfer the 1 mL to a serology tube.
3.  Add 3-5 drops of the reagent A and shake.
4.  Add 3-5 drops of the reagent B and shake.
5.  Wait at least 10 minutes.
6.  If a red color develops immediately, it is a positive test. Examine the Durham tube for gas formation. Result = positive nitrate reduction with or without $N_2$.
7.  If no red color develops, add some zinc dust. If a red color develops immediately, the result = negative nitrate reduction.
8.  If no red color develops after adding the zinc dust, the result = a positive nitrate reduction result. Examine the Durham tube for gas formation.
9.  Record all the results in the Evaluation of Results section.
10. Refer to the Evaluation of Nitrate broths reference table below.
11. If a negative test is obtained, then reincubate the original nitrate broth tube for another 24-48 hr. and retest as above.
12. Discard the serology tube with the reagents.
13. When the test is completed and confirmed negative or positive, discard all contaminated test tubes correctly.

**Evaluation of Nitrate Broths**

| Observations | Interpretations |
|---|---|
| No color change after the addition of reagents A and B. No gas in Durham tube. Medium-red after the addition of zinc. | Organism does not reduce nitrate to nitrite. No nitrogen gas ($N_2$). Nitrate is still present in the medium. |
| Medium-red after the addition of reagents A and B. No gas in Durham tube. | Organism(s) reduces nitrate to nitrite only; DO NOT ADD ZINC DUST. No nitrogen gas ($N_2$). |
| No color change after the addition of reagents A and B. No color change after the addition of zinc. Gas trapped in Durham tube. | Test organism is positive for denitrification: nitrate $\longrightarrow$ nitrite $\longrightarrow$ nitrogen Gas ($N_2$) All the nitrate and nitrite has been converted to nitrogen gas ($N_2$). |
| No color change after the addition of reagents A and B. No color change after the addition of zinc. No gas trapped in Durham tube. | No nitrate in medium, or the test organism is positive for denitrification (nitrate to nitrite to nitrogen gas) and the gas was not trapped in the Durham tube for some reason. Repeat the test. |

# J. OXIDASE

## MATERIALS

Safety: Biosafety Level 1

Cultures:

*E. coli* (negative) - BSL1
*P. stutzeri* (positive) - BSL1

Media:

TSA plate

Supplies:

Sterile applicator sticks
Oxidase test paper

## PROCEDURES
Day 1 (Inoculations)

*Technical Background*

The oxidase test is used for bacteria that contain cytochrome c as a respiratory enzyme. The test is used to separate the oxidase-positive Pseudomonadaceae (*Pseudomonas* species) from the oxidase-negative members of the Enterobacteriaceae. There are some variations of the way the test is performed, but the basic idea is to add the oxidase reagent to a colony of bacterium. If the reagent turns blue to black within 10 seconds, it is a positive test. If there is no color change, then the organism is oxidase-negative. A common way to perform the test is to use filter paper that is impregnated with the oxidase reagent. The bacteria should be freshly grown.

*Preparing Inoculations*

1. Aseptically transfer each of the organisms to one-half of a TSA plate.
2. Streak for pure isolated colonies.
3. Incubate the TSA plate at 35°C for 24-48 hr. (for fresh colonies).
4. Refrigerate plates if necessary until the next lab period.

## PROCEDURES
Day 2 (Test and Results)

*Evaluating Your Results*

1.  After growth has appeared:
    Use a sterile applicator stick to scrap some of the bacteria from the colony or growth on a slant and rub it on the oxidase paper. (Note: The sterile wood applicator stick is used instead of the nichrome loop, which might cause a false positive result.)
2.  The appearance of a blue-black color within 10 seconds is a positive test. No color change is a negative result

## K. UREASE TEST

## MATERIALS

Safety: Biosafety Level 1

Cultures:

*Citrobacter freundii* (positive) - BSL1
*E. coli* (negative) - BSL1

Media:

Urea slants

## PROCEDURES
Day 1 (Inoculations)

*Technical Background*

The purpose of the urease test is to determine the ability of an organism to split urea, forming two molecules of ammonia by the action of the enzyme urease with resulting alkalinity. The detection of urea hydrolysis can be very helpful in differentiating the genus *Proteus* from other members of the Enterobacteriaceae family, especially the pathogens. *Salmonella*, and *Shigella*, are lactose-negative, Gram-negative, enteric pathogens that are urease-negative. Proteus species can be confused with enteric pathogens because they do not ferment lactose. One way *Proteus* can be differentiated from the enteric pathogens is by a positive urease test.

*Preparing Inoculations*

1. Label the urea tubes appropriately.
2. Inoculate a loopful of the bacteria onto the slant, by streaking form the bottom of the slant to the top.
3. Incubate at 35ºC (loose caps) until the next lab period.

## PROCEDURES
Day 2 (Results)

*Evaluating Your Results*

1. Examine the urea tubes for growth.
2. Any appearance of a bright pink color is a positive test result.
3. No pink color or color change of the medium is a negative test.
4. Compare the inoculated tubes to uninoculated controls if necessary.
5. Record the results in the Evaluation of Results section.

# EVALUATION OF RESULTS
# (EXERCISE 17: SELECTED PHYSIOLOGICAL AND BIOCHEMICAL TESTS)

Purpose

Data

**Identification Tests**

|  | Test | Media and/or Reagents used | Enzyme involved | Appearance of positive result |
|---|---|---|---|---|
| A | Casein hydrolysis |  |  |  |
| B | Catalase test |  |  |  |
| C | Citrate utilization |  |  |  |
| D | Lactose (or other sugars) fermentation |  |  |  |
| E | Hydrogen sulfide production |  |  |  |
| F | Indole production |  |  |  |
| G | Mixed acid fermentation (MR) |  |  |  |
| H | Motility |  |  |  |
| I | Nitrate reduction |  |  |  |
| J | Oxidase test |  |  |  |
| K | Urea hydrolysis |  |  |  |

# EVALUATION OF RESULTS
## (EXERCISE 17: SELECTED PHYSIOLOGICAL AND BIOCHEMICAL TESTS)

Purpose

Data
**Results of Identification tests**

| Section | Test | Media and/or reagents used | E. coli | Citrobacter freundii | Serratia marscensens | E. aerogenes | Alkaligenes faecalis | Pseudomonas stutzeri | S. epidermidis | Bacillus subtilis | Streptococcus |
|---|---|---|---|---|---|---|---|---|---|---|---|---|
| A | Casein hydrolysis motility | | | | | | | | | | |
| B | Catalase test | | | | | | | | | | |
| C | Citrate utilization | | | | | | | | | | |
| D | Production of acid from lactose | | | | | | | | | | |
| E | Hydrogen sulfide production | | | | | | | | | | |
| F | Indole production | | | | | | | | | | |
| G | Mixed acid fermentation (MR) | | | | | | | | | | |
| H | Motility test | | | | | | | | | | |
| I | Nitrate reduction Nitrate to nitrite Nitrite to nitrogen gas | | | | | | | | | | |
| J | Oxidase test | | | | | | | | | | |
| K | Urea hydrolysis | | | | | | | | | | |

## CONCLUSIONS, DISCUSSIONS, AND QUESTIONS

1.  Discuss some difficulties one might encounter in differentiating bacteria on the basis of physiological tests.

2.  Hopefully, you noticed all the reminders to incubate the test tubes with loose caps. Why is this important?

3.  How would you determine whether an organism can ferment other carbohydrates, such as mannitol?

4.  What enzymes are involved in the following reactions:

    Urea hydrolysis–

    Hydrogen sulfide production–

    Casein hydrolysis–

    Indole production–

    Catalase–

5. Indicate the appearance of positive test results for:

Glucose fermentation with no gas–

Citrate utilization–

Urease test–

Indole production–

Hydrogen sulfide production–

Casein hydrolysis–

Catalase test–

Oxidase test–

# Medical Microbiology

The medical microbiology section introduces specific techniques used in the clinical laboratory. The student is expected to have learned the basic techniques from the first section and apply them to this section. Specific techniques are added as needed with each of the exercises. The Gram- negative rods were introduced at the end of the last section, as part of the biochemical tests exercise. The medically significant aspects of this group will be discussed in this section in various exercises. The section starts with the study of two clinically significant groups of bacteria: the Staphylococci and the Streptococci. Urine cultures will be introduced in this section, as well as antibiotic sensitivity studies. Soil will be used to try to isolate antibiotic-producing bacteria. Also, immunology techniques will be introduced. After learning the necessary techniques, you will be given unknown organisms to identify. The unknown exercise has been developed so that the organisms can be identified using the information from this manual.

Safety: Most of the laboratory exercises in this section will be performed using Risk Group 1 (RG1) and Risk Group 2 (RG2) microorganisms. This section will have certain laboratory exercises that will be performed with Risk Group 2 (RG2) microorganisms using special equipment and procedures designed for Biosafety Level 2 (BSL2). Students need to know the difference between Biosafety Level 1 and 2 microorganisms and should review all the rules for laboratory safety in the beginning of the manual.

# E X E R C I S E *18*

## Gram-Positive Cocci: Staphylococci

### OBJECTIVES

At the conclusion of the exercise, you should....

1. be able to name and recognize "normal flora" of the skin.
2. understand what a "carrier" is.
3. be able to discuss nosocomial infections in relation to the staphylococci.
4. understand opportunistic pathogen vs. pathogen.
5. recognize the Genus *Staphylococcus* by colony and slide morphology, as well as specialized tests.
6. understand the catalase test, coagulase test, DNAse test, and know how they are used in differentiating the Staphylococci.
7. recognize hemolysis on sheep blood agar.
8. observe and understand how the media SM 110 and mannitol salt agar are used in this exercise.

### INTRODUCTION

The Gram-positive cocci include a diverse number of Genera. Some examples are *Micrococcus, Enterococcus, Streptococcus,* and *Staphylococcus*. The staphylococci cells occur in clusters, singly, and in pairs. The opaque colonies can range from white to cream to yellowish, and they are **catalase positive**. They are mainly associated with the skin of warm-blooded vertebrates, but can be isolated from food products, dust, and water. Some species are **opportunistic pathogens** of humans and animals. Infections acquired while in the hospital are called nosocomial infections. These infections can result from several factors: 1) microorganisms in the hospital environment; 2) the compromised or weakened patient; and 3) a chain of transmission from hospital workers' normal flora. *Staphylococcus aureus* has been involved in many nosocomial infections. In this exercise, you will try to isolate *Staphylococcus aureus* and *S. epidermidis* from your nose and skin, as well as a fomite (inanimate object). You will observe some of the various biochemical reactions for this group of bacteria by performing the tests on the known species provided and on the "unknown" organisms recovered from the fomite and nose.

## MATERIALS

Safety: Biosafety Levels 1 and 2

Cultures:

*Staphylococcus aureus* (*S. aureus*) TSB - BSL2
*Staphylococcus epidermidis* (*S. epi*) TSB - BSL1
Cultures obtained from nose and fomite - BSL1 & 2

Media:

M-staphylococcus broths
mannitol salt agar plates
SM 110 plates
Sheep blood agar plates (BAP)
Coagulase plasma (rabbit) test tubes
DNA agar plate

Supplies:

Sterile swabs
Hydrogen peroxide (3%) per room (fresh)
Water bath set at 37°C

## PROCEDURES
### Day 1 (Collection and Enrichment)

*Technical Background*

M-Staphylococcus broth is a selective medium that contains sodium chloride, which will restrict the growth of many bacteria and, thus, select for the Staphylococci.

*Collecting M-staphylococcus Samples*

1.  One student should obtain a sample from his or her own nose with a sterile swab. Make sure the swab is inserted into the back of the nostril (called anterior nare), and also touch some of the skin around the opening. Then, place the swab in the M-staphylococcus broth labeled "Nose" (Note: Swirl the swab in the broth and break off the top of the stick if the cap does not go back on the tube.)
2.  Another student will obtain a sample from a fomite in the lab or from the surrounding area. This could be the lab bench, the bottom of your shoe, the handle on the paper towel dispenser, the hallway, etc. Make a note of what area was sampled. Place the swab in the tube of M-staphylococcus broth, labeled "Fomite," using the same technique as above in Step 2.

3. Using aseptic technique, transfer a loopful of *S. epidermidis* to the M-staphylococcus broth labeled *S. epi.*

4. Using aseptic technique again, transfer a loopful of *S. aureus* to the M-staphylococcus broth tube labeled *S. aureus.*

5. Place all 4 tubes into a beaker or test tube rack labeled with the date, exercise #, your section #, and table #.

6. Incubate the tubes of M-staphylococcus broth at 35°C, until the next lab period.

# PROCEDURES
## Day 2 (Isolation)

### Technical Background

MSA, Mannitol Salt Agar, is a selective and differential type of medium that has an NaCl salt concentration of 7.5%. It selects for the isolation of staphylococci, since they can grow on high concentrations of NaCl. Gram-negative and some other bacteria should be inhibited because of the high concentration of salt.

> *S. aureus* produces colonies with a yellow zone that is due to production of acid from mannitol. *S. epidermidis* colonies produce no color change of the medium; therefore, the media surrounding them appears red.

Staphylococcus medium 110 (SM 110) is another selective medium for the isolation of staphylococci. Again, it is the high concentration of NaCl that makes it selective for any staphylococci.

> *S. aureus* produces a yellow pigment on SM 110.
> *S. epidermidis* is negative for pigment production on SM 110; therefore, it appears white.

### Preparing Inoculations

1. Label one-half of an MSA and SM 110 plate with the known organism *S. epidermidis* and the other half with *S. aureus.*

2. Label the second MSA and SM 110 plates "Nose."

3. Label the third MSA and SM 110 plates "Fomite."

4. Transfer a loopful of the known *Staphylococcus* species to the correctly labeled half of the MSA and SM 110. Streak each half for isolation.

5. Transfer a loopful from the M-staphylococcus broth labeled Unknown-Fomite to the correctly labeled MSA plate, and repeat with another loopful to the SM 110 plate. Streak both plates for isolation.

6. Repeat the above inoculations with the Unknown-Nose sample.

7. Incubate all plates at 35°C, until the next lab period.

## PROCEDURES
### Day 3 (Observations and Identification Tests)

*Technical Background*

**Coagulase Test:** Once an organism is isolated from a patient specimen and identified as a Gram-positive, catalase-positive cocci, the next step is to perform a coagulase test. Coagulase is an enzyme, which binds plasma fibrinogen, causing plasma to clot. This test can distinguish *S. aureus* from other staphylococci. *S. aureus* produces the coagulase enzyme; therefore, it will clot the rabbit plasma in 1-4 hours.

> *S. aureus* = positive (plasma clotted)
> *S. epi* = negative (plasma liquid)

**DNAse Test:** This test aids in further identification of *S. aureus*, which produces Deoxyribonuclease, an enzyme that can alter DNA. When a culture of *S. aureus* is inoculated onto a plate of DNAse agar with toluidine blue, the nuclease cleaves the DNA, which changes the color of the agar from blue to pink. *S. epidermidis* can degrade the dye without cleaving the DNA, thus causing a clear zone.

> *S. aureus* = positive (pink halo)
> *S. epidermidis (S. epi.)* = negative (clear halo)

**Catalase Test:** See description in Biochemical and Physiological Tests, for the technique. This test is primarily used to differentiate between genera. In this case, you will perform the catalase test on Gram-positive cocci, to determine if they are in the genus *Staphylococcus* or *Streptococcus*.

> *Staphylococcus* = catalase positive
> *Streptococcus* = catalase negative

**Types of hemolysis:**
Blood Agar Plate, 5% Sheep Blood (BAP), is a differential-type media used as a general nutritional growth medium for the cultivation of most bacteria, as well as to determine hemolytic reactions (the lysis of red blood cells). Determining the type of hemolysis produced by bacteria often is used as the initial screening to assist in determining what other steps should be taken for the identification of an isolate.

Alpha (α) – incomplete or partial lysis of the red blood cells: causes a greenish darkening of the
     agar
Beta (ß) – complete lysis of the red cells: causes a transparent (or clearing) of the agar
Gamma (δ) – no lysis of the red cells: no change in the blood agar
*Staphylococcus aureus* produces hemolysin, an enzyme that causes beta (complete) hemolysis
*Staphylococcus epidermidis* does not produce hemolysin and is non-hemolytic

> *S. aureus* = Beta hemolytic
> *S. epi* = Non-hemolytic

## Preparing the Inoculations

1. Examine the MSA and SM 110 plates with *S. epidermidis* and *S. aureus*, and note the color of the colonies. They should appear as described above.
2. Perform a catalase test on both to confirm that they are Staphylococci.
3. Look for similar colonies from the nose and fomite plates. Choose one that appears to be *S. epidermidis* and another that appears to be *S. aureus*. Do a Gram stain and catalase test on these colonies, also using one of your QC slides to be certain the procedure is accurate.
4. Perform the coagulase test on each of the above colonies (*S. aureus, S. epidermidis,* unknown from nose and unknown from fomite) that are **catalase positive**. (Note: Use a heavy inoculum.)
5. Incubate the coagulase tubes in a 35°C incubator until the next lab period.
6. Label the four BAPs, *S. aureus, S. epi.,* Nose, and Fomite. Inoculate each, streaking for isolation.
7. Divide a DNAse plate into four sections and inoculate each section with each of the above colonies. The inoculum should be inoculated as a spot the size of a dime and very heavy. See diagram:

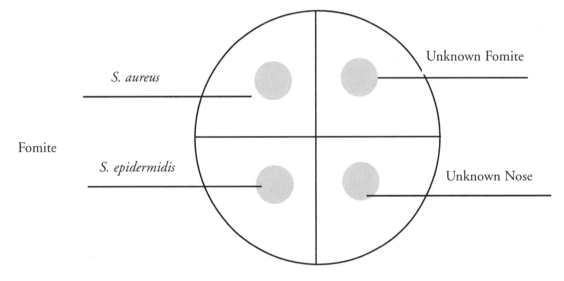

8. Incubate all plates at 35°C until the next lab period.
9. Record all results so far in the Evaluation of Results table.

## PROCEDURES
Day 4 (Observations and Identification, Cont'd.)

*Evaluating Your Results*

1.  Read and record the coagulase tube results by tilting the test tube and looking for coagulation of the rabbit plasma. (Note: *S. aureus* is coagulase positive and *S. epidermidis* is coagulase negative.)
2.  Read and record the DNAse test results. (Note: You are looking for a pink halo surrounding the inoculum (positive result) or a clear halo (negative result)).
3.  Read and record the BAP plate hemolysis results. (Note: Most *S. aureus* are Beta hemolytic and *S. epidermidis* are non-hemolytic.)
4.  Record all results in the Evaluation of Results section.

## EVALUATION OF RESULTS
## (EXERCISE 18: GRAM-POSITIVE COCCI: STAPHYLOCOCCI)

Purpose

Data

| Test | S. aureus | S. epi | Nose | Fomite |
|------|-----------|--------|------|--------|
| **BAP plate**: color of colonies | | | | |
| type of hemolysis | | | | |
| **MSA plate**: color of zone surrounding colonies | | | | |
| **SM 110 plate**: color of colonies | | | | |
| **DNAse plate**: color of zone surrounding colony | | | | |
| **Catalase test reaction** | | | | |
| **Coagulase test reaction** | | | | |
| **Gram stain reaction & shape and cell arrangement** | | | | |

## CONCLUSIONS, DISCUSSIONS, AND QUESTIONS

1. Discuss how the coagulase test is a characteristic for pathogenicity.

2. What are nosocomial infections?

3. Why were mostly Gram-positive cocci isolated from the fomite and nose cultures, when a variety of organisms could have been present at these sites?

4. Define the following:

   Normal flora–

   Carrier–

5. *Staphylococcus aureus* contains virulence factors that can be demonstrated in the laboratory. List two, and describe the effects displayed by the testing procedure.

6. *S. aureus* is responsible for several different clinical syndromes. Using your textbook and notes from lecture and lab, list five that are found with *S. aureus* as a principle cause.

7. Compare *S. aureus* with *S. epidermidis* with respect to their morphology and the biochemical tests that differentiate them.

# EXERCISE 19

## Gram-Positive Cocci: Streptococci

### OBJECTIVES

At the conclusion of the exercise, you should...

1. be able to recognize the three different types of hemolysis on sheep blood agar.
2. be able to list normal microbial flora of the mouth and respiratory tract.
3. know how to differentiate the streptococci, based on colony morphology and biochemical testing.
4. be able to recognize streptococcal chains with a Gram stain.
5. know how to perform the catalase test to aid in classifying the streptococci.
6. learn the technique for obtaining a throat culture.
7. understand the use of identification discs.
8. learn what streptococci are responsible for dental plaque and dextran or levan production.

### INTRODUCTION

The streptococci are **catalase-negative**, **Gram-positive cocci** that tend to grow in chains in liquid media. They can be identified in the laboratory by several different methods. The clinical microbiology laboratory uses identification methods that involve preliminary grouping of the isolates based on morphology and hemolysis of colonies growing on sheep blood agar. In this exercise, you will collect a throat culture from your own throat and identify the streptococci from it, using selected clinical microbiological methods.

### MATERIALS

Safety: Biosafety Levels 1 and 2

Cultures:

Blood agar plates showing hemolysis and Todd Hewitt broths as needed of the following:
    *Enterococcus faecalis* (Group D, Alpha/gamma hemolytic) - BSL2
    *Streptococcus viridans* (Viridans group, Alpha hemolytic) - BSL2
    *Streptococcus pyogenes* (Group A, Beta hemolytic) - BSL2
    *Streptococcus salivarius* (Viridans group, Alpha/gamma hemolytic) - BSL1
    *Streptococcus pneumoniae* (Alpha hemolytic) - BSL2
    *Streptococcus equi* (Group C, Beta hemolytic) - BSL2

Media:

BAP plates
Todd Hewitt (TH) broths
6.5% NaCl broths
Bile esculin slants

Supplies:

Sterile swabs
Tongue depressors
Candle jars
Bacitracin (A) disks
Optochin (P) disks
SXT disks: 1.25 mg trimethoprim + 27.75 mg sulfamethoxazole

## PROCEDURES
### Day 1 (Throat Cultures)

*Technical Background*

The respiratory tract can be divided into the upper and lower respiratory systems. The upper consists of the nose and throat. The lower respiratory system consists of the larynx, trachea, bronchial tubes, and alveoli. The lower is normally a sterile environment, whereas the upper system is in contact with the air we breathe that is contaminated with microorganisms. The throat is a moist, warm environment, and many different genera of bacteria reside there. These include *Corynebacterium, Hemophilus, Neisseria, Staphylococcus,* and *Streptococcus.* The streptococci are the most predominant microbes found in the throat. The mouth contains millions of bacteria. Some are carried into the mouth on food, and some species of *Streptococcus* are part of the normal microbiota of the mouth. One example is *S. salivarius. S. salivarius* is one of the organisms that is responsible for dental plaque and can produce the dextran or levan capsule, which enables the bacteria to adhere to the teeth, thus causing cavities. Dextran is a sticky polysaccharide produced from glucose, and levan is formed from sucrose.

The growth of many streptococcal isolates is stimulated in a $CO_2$-enriched environment. When $CO_2$ incubators are not available, a candle jar is used.

**$CO_2$ or Candle jars** are used to culture organisms that require carbon dioxide for growth. The candle jar is made from any container that can hold several Petri dishes (or a beaker with test tubes) and can be tightly sealed. After placing inverted Petri plates into the candle jar, a candle is set in the jar on top of the inoculated plates. Some of the oxygen inside the jar is used up by the burning candle and replaced by carbon dioxide. When the candle goes out, the jar contains about 5% carbon dioxide. The candle jar does not produce an anaerobic environment.

*Preparing the Inoculation*

1. Use a sterile cotton swab to swab your own throat. Use a mirror and tongue depressor if necessary.
2. Swab the area on both sides of the uvula (see diagram).
3. Label the bottom of a blood agar plate (BAP) and divide into sections for the T-streak technique.
4. After obtaining the inoculum from the throat, swab approximately one-third to one-half of a blood agar plate (BAP) for the area of primary isolation.
5. Streak the remaining sections of the plate with a sterile loop. Streak for isolation as described in the T-streak technique.
6. Invert the BAP plate and place it in the candle jar.
7. Incubate the candle jars for 24 hours at 35ºC.
8. After 24 hours, move the candle jars to a refrigerator until the next lab period.

### THROAT SWAB AND CANDLE JAR PROCEDURES

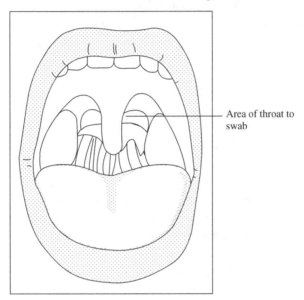

Area of throat to swab

Light candle.

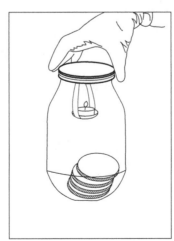

Replace lid with candle still lit, and close lid slowly so candle continues to burn.

Allow candle to burn out and place the candle jar into 35°C incubator.

## PROCEDURES
### Day 2 (Enrichment)

*Technical Background*

There are three types of hemolysis that are used to initially classify the streptococci:

α (alpha) is **incomplete** hemolysis, producing a **green**, cloudy **zone** around the colony.

β (beta) is **complete** hemolysis, producing a **clear zone** around the colony.

γ (gamma) is **no** hemolysis, and **no change** in the blood agar around the colony.

The second major criteria for identification of the streptococci is the antigen group present on the cell wall. These antigens were originally defined by Rebecca Lancefield and are called the Lancefield groups, A through O. The technique for testing for these groups is covered in a subsequent section.

The streptococci that are alpha and gamma hemolytic are usually normal flora, whereas beta hemolytic streptococci are frequently pathogens. Most beta hemolytic streptococci that are pathogenic to humans belong to **Group A**, and are classified as *Streptococcus pyogenes*.

*Classifying the Streptococci*

1.  Examine the known species of streptococci for the different types of hemolysis.
2.  Perform a catalase test on a positive control and a negative control.
3.  Examine your own throat culture plate.
4.  Choose a colony that is alpha hemolytic and perform a catalase test.
5.  Choose a colony that is gamma hemolytic and perform a catalase test.
6.  Choose a colony that is beta hemolytic and perform a catalase test.
7.  Most throat cultures will have alpha and non-hemolytic, catalase-negative colonies growing. After identifying an alpha and/or gamma hemolytic streptococcus colony, perform Step 9.
8.  Between you and your lab partner, select one or a few of the catalase-negative, Gram-positive cocci colonies and transfer them into a Todd Hewitt broth for enrichment. (Note: If you have a catalase-negative Beta hemolytic colony, then choose that colony to enrich.)
9.  Place the TH tube into a beaker that will fit into a candle jar.
10.  Light the candle and close the lid of the candle jar.
11.  After the flame goes out, place the candle jar in the 35°C incubator until the next lab period.
12.  Record all results in the Evaluation of Results data table.

# PROCEDURES
## Day 3 (Identification Tests)

*Technical Background*

Identification methods used to differentiate the streptococci include bacitracin susceptibility, Optochin susceptibility, bile esculin hydrolysis, 6.5% salt tolerance, and immuno serological identification. (See table below.)

**Bacitracin (A disc)** susceptibility test is used for the **presumptive** identification of group **A beta hemolytic streptococci** on blood agar (**Streptococcus pyogenes**). *S. pyogenes* is inhibited (sensitive) by a 0.04 U bacitracin disk.

**SXT** disc susceptibility: 1.25 mg trimethoprim plus 27.75 mg sulfamethoxazole is used for the presumptive identification of beta hemolytic streptococci on blood agar. Group A beta hemolytic strep is susceptible to the A disc, but resistant to SXT. Group B beta hemolytic strep is resistant to both A and SXT discs. Other types of beta-hemolytic streps are resistant to the A disc, but sensitive to SXT (example Group C). See Appendix E: Unknown identification table: Gram-positive cocci – catalase-negative.

**Optochin (P disk)** susceptibility is used to differentiate between **Streptococcus pneumoniae** and other alpha hemolytic streptococci. *S. pneumoniae* growth is inhibited around the **P disc**.

**Bile esculin hydrolysis** is used to aid in the differentiation of **Group D streptococci, enterococci** from other streptococci, not Group D. **Enterococcus faecalis**, a **Group D** streptococcus, can grow in the presence of 40% bile and can hydrolyze **esculin to esculetin**. Esculetin forms a **black complex** with the ferric citrate in the bile esculin agar (**bile esculin positive**). Another test for *E. faecalis* is growth in **6.5% NaCl broth**. Group D enterococci are able to grow in 6.5% NaCl broth.

*Identifying the Streptococci*

1. Examine the demo BAP plates of the known streptococci.
2. Make sure you know how to recognize the three types of hemolysis, and record the hemolysis of each of the known streptococci in the Evaluation of Results data table. Follow the summary diagram below, and if the assigned known strep is
   Beta hemolytic, then transfer a loopful from the TH broth to the BAP: streak the BAP plate for isolation and place an A disc and SXT disc onto the primary streak area (see diagram).

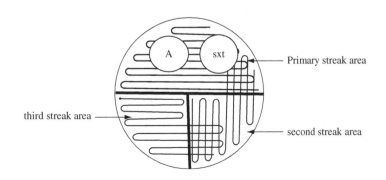

<u>Alpha or non-hemolytic</u>, then transfer a loopful from the TH broth to the BAP: streak a BAP plate for isolation and place a P disc onto the primary streak area (see diagram) and inoculate a bile esculin slant and a 6.5% NaCl broth tube (loopful each).

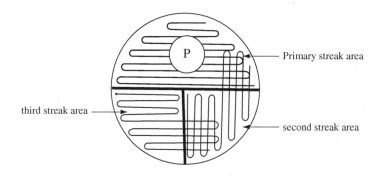

3.  Repeat the same procedures above, based on the type of hemolysis for the "unknown" colony that was picked from your throat and enriched in TH.
4.  Invert all the BAP plates, and incubate in the candle jar at 35ºC for 24 hours.
5.  After 24-hr. incubation, move the candle jar to a refrigerator until the next lab period.
6.  Incubate the bile esculin slants and 6.5% NaCl tubes at 35ºC in the air incubator until the next lab.

**Summary of how to choose colonies from the throat cultures:**

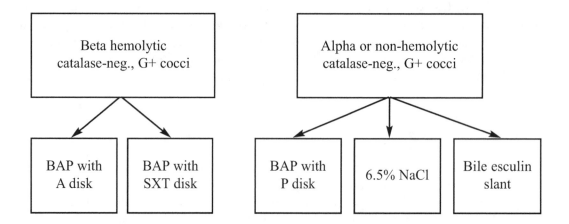

# PROCEDURES
## Day 4 (Evaluation of Results)

*Technical Background*

Table 1. Differentiation of the Streptococci.

| Tests | Group A S. pyogenes | Group B S. agalactiae | Group C S. equi | Group D enterococci E. faecalis | Group D non-enterococci S. bovis | Oral Strep S. mutans S. salivarius | S. pneumoniae |
|---|---|---|---|---|---|---|---|
| Type of Hemolysis | beta (β) | beta (β) | beta (β) | alpha (α) or gamma (γ) (none) | alpha (α) or gamma (γ) (none) | alpha (α) or gamma (γ) (none) | alpha (α) |
| Bacitracin (A) disc Sensitivity | S | R | R | ND | ND | ND | ND |
| SXT | R | R | S | ND | ND | ND | ND |
| Optochin (P) disc Sensitivity | ND | ND | ND | R | R | R | S |
| Bile esculin hydrolysis | ND | ND | ND | Positive | Positive | Negative | Negative |
| 6.5% NaCl tolerance (growth in) | ND | ND | ND | Positive | Negative | Negative | Negative |

Key:

| | |
|---|---|
| Negative | negative growth or hydrolysis |
| Positive | positive growth or positive hydrolysis |
| S | sensitive to drug |
| R | resistant to drug |
| ND | not done |

1.  Use the above table as a guide to evaluate the throat cultures and the known cultures.
2.  Make sure you see and can interpret all of the known streptococci identification tests.
3.  Read and record all of your results and your partner's results in Evaluation of Results.

# EVALUATION OF RESULTS
## (EXERCISE 19: GRAM-POSITIVE COCCI: STREPTOCOCCI)

Purpose

Data

| Specimen | Morphology | Cell Arrangement | Hemolysis | Gram Stain | Catalase Test | A disc | SXT Disc | P Disc | Bile Esculin | 6.5% NaCl | Presumptive Id |
|---|---|---|---|---|---|---|---|---|---|---|---|
| **Throat:** Colony 1 | | | | | | | | | | | |
| Colony 2 | | | | | | | | | | | |
| Colony 3 | | | | | | | | | | | |
| Colony 4 | | | | | | | | | | | |
| **Known:** E. faecalis | | | | | | | | | | | |
| S. viridans | | | | | | | | | | | |
| S. pneumoniae | | | | | | | | | | | |
| S. pyogenes | | | | | | | | | | | |
| S. salivarius | | | | | | | | | | | |
| S. equi | | | | | | | | | | | |

## CONCLUSIONS, DISCUSSIONS, AND QUESTIONS

1.  What two tests can be used to distinguish *Streptococcus pyogenes* from other members of the Streptococci?

2.  List three diseases that Group A *Streptococci* cause in humans.

3.  What test can be used to identify *Streptococcus* pneumoniae from other members of the Streptococci?

4.  Explain the connection between a diet rich in sucrose and tooth decay. (Hint: Sucrose consists of glucose and fructose linked together. It is the chemical we buy as "sugar"; it comes from sugar cane or from sugar beets.)

# E X E R C I S E 20

## Urine Cultures

## OBJECTIVES

At the conclusion of the exercise, you should...

1. know how to collect a clean-catch urine sample.
2. be able to use universal precautions when handling body fluids.
3. be able to determine the presence or absence of a urinary tract infection.
4. be able to name the most common species of bacteria that cause a urinary tract infection.

## INTRODUCTION

The **urinary tract** consists of the kidneys, the ureters, the bladder, and the urethra. In healthy people, only the urethra is colonized by bacteria. But all portions of the urinary tract are susceptible to chronic or acute infections caused by many different bacteria. *E. coli* is the most common culprit, followed by *Proteus spp*,-*Pseudomonas* spp., enterococci, streptococci, and staphylococci. During the infection, the pathogen is usually voided in the urine, along with leucocytes and, sometimes, erythrocytes. But the mere presence of bacteria in the urine is not diagnostic of a **urinary tract infection (UTI)** because the urethra hosts such resident microbiota as Gram-positive rods (corynebacteria) and cocci (streptococci), and Gram-negative cocci (Neisseria spp.), and such transient microbes as enteric bacteria (Gram-negative rods) and yeasts. All of these may be shed in the urine. Because the non-invasive method for collecting urine relies on urine that has passed through the colonized urethra, special steps must be taken to minimize contamination of the urine during collection. Also, it is important to determine the number of microbes per mL of urine. The colonizers are usually present in urine at numbers less than 10,000/mL, whereas numbers of bacteria greater than 10,000 to 100,000/mL suggest an infection, especially if one colony type appears on the agar plates.

In this exercise, you will learn the technique for collecting a clean-catch urine sample, and how to test it for evidence of UTI. Each table will analyze two urine specimens: (A) a freshly collected specimen from one (1) person at the table, and (B) an "infected" artificial urine. Throughout the exercise, you will follow the universal precautions in handling body fluids.

## MATERIALS

Safety: Biosafety Levels 1 and 2

Media:

TSA plates

Supplies:

Container of "infected" urine
Mid-stream Urine Collection Kit
Packages of sterile 1 µl (1 microliter) loops
Packages of sterile 10 µl (10 microliter) loops
Multistix® 10 SG reagent strips for urinalysis
Color chart for reading the reagent strips

# PROCEDURES
## Day 1 (Urine Collection, Dipstick Testing, and Plating)

*Technical Background*

A clinically relevant urine specimen must be collected in such a way that the normal vaginal, perineal, and anterior urethral microbiota do not contaminate it. The least invasive procedure is the **clean-catch, midstream urine specimen**. More invasive procedures include catheterization to collect urine directly from the bladder or suprapubic bladder aspiration using a syringe and needle; these invasive procedures risk introducing pathogens into the bladder. Because bacteria can grow in urine, it is important to deliver the urine specimen to the laboratory soon after collection; the laboratory should initiate testing within 1 hour of collection. Refrigeration can extend these times.

The Dipstick test is a quick way to rapidly screen urine specimens for signs of **UTI (urinary tract infection)**; in addition, the Dipstick supplies information on the overall health of the individual. Each Dipstick has several pads, one pad for each test. The Dipstick is lowered into the urine, removed, and the color on each pad compared with the color chart.

When the Dipstick test indicates a UTI, or if the patient has symptoms (pain, frequency, urgency of urination), a microscopic or culture-based examination of the urine is ordered. The number of culturable bacteria per mL of urine aids in the differential diagnosis of a UTI. Typically, a very small, but known, volume of the urine is spread out on an agar plate and, after incubation, the colonies are counted; conversion to number of colonies/mL is made after incubation. Greater than 100,000 colonies/mL of one probable pathogen indicates a urinary tract infection; fewer bacteria/mL may be considered significant if the patient has symptoms. If there are small numbers of a variety of colony types, the urine was probably contaminated by normal microbiota and, if the clinician suspects a UTI, a new specimen will be obtained.

**Universal precautions for prevention of transmission of HIV and other bloodborne infections** were published by the Centers for Disease Control (**CDC**) in 1987. Under these precautions, blood and certain body fluids of ALL patients are considered potentially infectious. The precautions apply to blood and other body fluids containing visible blood, semen, and vaginal secretions. Universal precautions also apply to tissues and cerebrospinal, synovial, pleural, peritoneal, pericardial, and amniotic fluids. They do not apply to feces, nasal secretions, sputum, sweat, tears, urine, saliva, and vomitus, unless visible blood is present. (Exception: In the dental setting, blood contamination of saliva is common so the precautions must be followed.) In this exercise, you will gain experience with the universal precautions by following them in the handling of urine. For more information, access the CDC website: www.cdc.gov/ncidod/hip/Blood/UNIVERSA.HTM.

*Following Universal Precautions for this Exercise*

- Wear gloves when handling urine. If you are sensitive to latex, request vinyl gloves.
- Wear your lab coat and closed-toed shoes.
- Use a protective barrier to protect the mucous membranes of your mouth and eyes. In this lab, the barrier will be a plexiglass shield placed on the lab bench. Pour, test, and pipette the urine from behind the shield.

*Collecting a Clean-Catch Urine Specimen*

1. Follow the directions that come with the Mid-Stream Urine Collection Kit. The volunteer from each table goes to the restroom.
2. Wash hands with soap.
3. Remove towelette.
4. **Females**: Separate the folds of the labia with the thumb and forefinger and clean inside with the towelette, using downward strokes only. Keep the folds separated during the urination into the container.
   **Males**: Clean the head of the penis with the towelette. Remove the container and DO NOT touch the inside of the container.
5. Begin urination into the toilet. As urination continues, bring the container into the stream of urine.
   a. Fill the specimen container only halfway.
   b. Remove the cap from the package. Do Not Touch the inside of the cap.
   c. Screw the cap on the container.
6. WASH YOUR HANDS and the outside of the container.
7. Place the container back into the plastic wrap and carry it to the lab.

*Set Up Universal Precautions at Your Lab Bench*
(work as pairs)

1. Wash your hands BEFORE and AFTER doing any lab work.
2. Disinfect the work area BEFORE and AFTER working with any specimens.
3. Wear gloves when handling the urine. If you are sensitive to latex, request vinyl gloves.
4. Set the protective barrier (plexiglass shield) on towels soaked with disinfectant.
   a. When working with urine, all students will work behind the shields to protect themselves and other laboratory workers near them.
   b. All students will wear gloves.
5. After finishing all parts of the lab exercise, carefully carry the urine back to the restroom wrapped in the original package and FLUSH THE URINE ONLY!
   a. Bring the urine cup and wrapper back to the classroom.
6. Dispose of the paper portion in the regular trash.
7. Dispose of the urine cup in the red Biohazard trash.
8. Please **Do Not** leave any urine materials in the restroom.
9. WASH YOUR HANDS.

## EVALUATION OF RESULTS
## (EXERCISE 20: URINE ANALYSIS)

Purpose

Data: Results of Urine Dipstick Testing:
(Work behind the plexiglass shield over disinfectant-soaked towels.)
  With the cap tight, swirl the urine in the cup.
  Have your partner prepare to record the results as you call them out (see data sheet). All
  tests must be read within 2 minutes of dipping the test strip into the urine.
  Obtain one (1) dipstick, a Kimwipe, and the package from the instructor.
  • Dip the stick into the urine and immediately remove it.
  • VERY IMPORTANT: Run the back of the stick along the rim of the cup and wipe
    the back of the dipstick with a Kimwipe.
  • Compare the color of the Glucose pad (#1) to the chart on the package. Make the
    comparison 30 seconds after dipping. Record results.
  • Then move to the next pad, etc. Time your readings accurately.

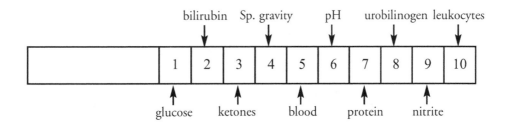

Table 1. Dipstick Tests (those important in diagnosing UTI in **bold**).

| Pad | Test name | Read in | What it detects | Abnormal test indicates |
|---|---|---|---|---|
| 1 | Glucose | 30 sec. | Glucose in urine. | Possible diabetes. |
| 2 | Bilirubin | 30 sec. | Degradation product of hemoglobin. | Liver or gall bladder problems. |
| 3 | Ketones | 40 sec. | End products of rapid or excessive fatty acid breakdown. | Uncontrolled diabetes, starvation or fasting, anorexia, high-protein or low-carbohydrate diet, protracted vomiting. |
| 4 | Specific gravity | 45 sec. | The concentration of solutes in the urine. Normal range: 1.002 to 1.028. | Less than 1.002, patient is drinking excessive fluid, as in diabetes. More than 1.028, the patient is dehydrated. |
| 5 | **Blood** | 60 sec. | Hemoglobin or intact erythrocytes. | Early sign of kidney or urinary tract disease. |
| 6 | **pH** | 60 sec. | pH of urine should be 4.6-8. | Alkaline pH may indicate UTI (bacterial urease splitting urea to ammonia), renal failure, vomiting. Acidic pH may indicate emphysema, diabetic ketoacidosis, and diarrhea. |
| 7 | **Urine protein** | 60 sec. | Urine normally contains 0 to 8 mg protein/100 mL. | Higher concentrations of protein may indicate a host of problems, e.g., congestive heart failure, kidney infection. |
| 8 | Urobilinogen | 60 sec. | Normal: 0.2-1 mg/dl | Increased level is an early sign of liver disease and hemolytic disorders. |
| 9 | **Nitrite** | 60 sec. | Presence of nitrite in urine. | 90% of UTI are caused by *E. coli*, which re-dukes nitrate to nitrite. Presence of nitrite in urine indicates UTI. If test is negative, but patient has UTI symptoms, then urine should be cultured for Gram-positive bacteria. |
| 10 | **Leukocytes** | 2 min. | Normal urine has no leukocytes. If present, they release esterases that turn the pad purple. | Indication of UTI or kidney infection. |

## URINE DIPSTICK PROCEDURE

Urine set up behind shield.

Remove Dipstick from jar.

Dip the Dipstick into urine.

Blot the Dipstick onto paper towel.

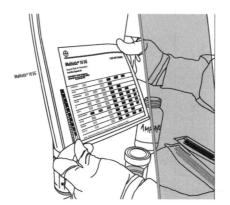

Read the Dipstick by comparing to color chart. Record results in Evaluation of Results.

## URINE CULTURE PROCEDURE

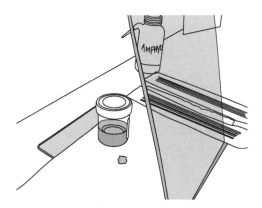

Urine and loops set up behind shield.

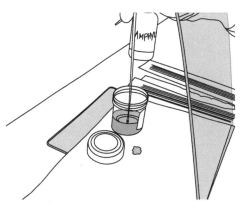

Dip 1µl loop into urine cup.

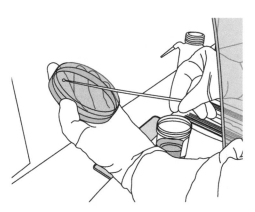

Streak plate (see diagram).

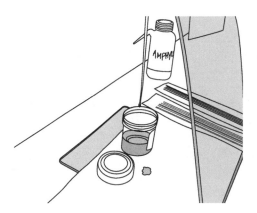

Dip 10 µl loop into urine cup.

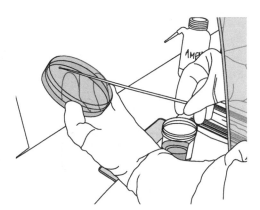

Streak plate (see diagram).

Place loop in SHARPS container.

## Dilution Plating

(Note: Do all work behind the plexiglass shield over disinfectant-soaked towels.) One pair works with the clean-catch urine. The other pair works with the "infected" urine. Directions are for each pair.

1.  Label one TSA plate 0.001 mL (1 μl).
2.  Label one TSA plate 0.01 mL (10 μl).
3.  Swirl the urine cup gently to mix the contents.
4.  Aseptically remove a sterile 1 μl loop from the package.
5.  Hold the loop in a vertical position and dip into the urine, just past the loop.
6.  Carefully remove the loop and streak the 0.001-mL, labeled, TSA plate down the middle of the agar once.
7.  Using the same loop, cross-streak back and forth over the first streak, covering the entire plate (about 20-30 times). See the diagram below.
8.  Discard the disposable loop in the SHARPS container, since it could puncture the red bag.
9.  Repeat the same procedure with a 10 μl loop onto the 0.01-mL, labeled, TSA plate.
10. Invert the agar plates and incubate at 35°C until the next lab period.

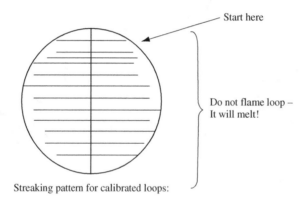

Streaking pattern for calibrated loops:

SUMMARY

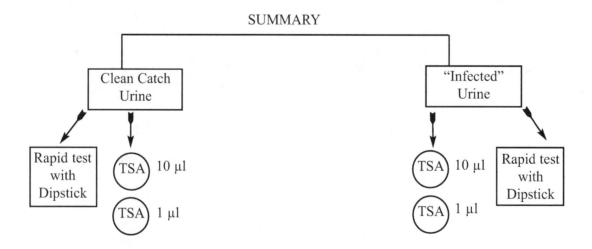

## PROCEDURES
## Day 2 (Plate Counts)

*Technical Background*

Most bacteria that are responsible for UTIs can be placed into two groups: Gram-negative rods and Gram-positive cocci. The Gram-negative rods cause most UTIs; these are *E. coli, Proteus* spp., *Enterobacter* spp., and *Klebsiella* spp. These bacteria will grow well on TSA, blood agar (BA), and EMB agar. **EMB (eosin methylene blue) agar** is a special agar used to isolate and differentiate Gram-negative bacteria. Most Gram-positive bacteria will not grow on EMB. It contains lactose, so if the bacteria ferment this sugar and make acid, the colonies turn purple. If the bacteria cannot ferment lactose, the colonies are colorless.

**BAP (5% sheep blood)** is, like TSA, a general-purpose agar that will grow Gram-negative and Gram-positive bacteria. When a UTI is caused by a Gram-positive bacterium, the most likely genera are *Streptococcus, Staphylococcus,* or *Enterococcus.* You have learned how to distinguish these genera using morphology and cell arrangement, hemolytic reaction on BA, and other characteristics.

When the TSA plates streaked with a known amount of urine show more than 100,000 bacteria per mL in clean-catch urine, this is evidence of a urinary tract infection. Symptoms of the UTI are nearly always present. A count of between 10,000 and 100,000 bacteria per mL is often considered evidence of a UTI if the patient has one or more symptoms.

1. Count the number of colonies on each of the TSA dilution plates.
2. Enter the counts under Quantitative Results on the next page.
3. Calculate the number of bacteria/mL of urine using the formula:

Number bacteria per mL = Number of colonies counted/amount plated in mL

Examples:

If 2 colonies grow on the plate inoculated with the 10 µl loop plate, then
    Number bacteria/mL = 2 colonies/0.01 mL = 200 bacteria/mL of urine.

If 3 colonies grow on the plate inoculated with the 1 µl loop plate, then
    Number bacteria/mL = 3 colonies/0.001 = 3000 bacteria/mL of urine.

## EVALUATION OF RESULTS
## (EXERCISE 20: URINE CULTURES)

Purpose

Data

Table 1. Urine Dipstick Test Results.

| Pad | Test name | Read in | Result 1 | Result 2 |
|---|---|---|---|---|
| 1 | Glucose | 30 sec. | | |
| 2 | Bilirubin | 30 sec. | | |
| 3 | Ketones | 40 sec. | | |
| 4 | Specific gravity | 45 sec. | | |
| 5 | **Blood** | 60 sec. | | |
| 6 | **pH** | 60 sec. | | |
| 7 | **Urine protein** | 60 sec. | | |
| 8 | Urobilinogen | 60 sec. | | |
| 9 | **Nitrite** | 60 sec. | | |
| 10 | **Leukocytes** | 2 min. | | |

**Quantitative Results:**

Number of colonies on the 0.001-mL TSA plate: _____

Dilution: _____

Number of organisms/mL of urine: _____

Number of colonies on the 0.01-mL TSA plate: _____

Dilution: _____

Number of organisms/mL of urine: _____

Show calculations below:

# CONCLUSIONS, DISCUSSIONS, AND QUESTIONS

1. What would you report to the attending physician after the Dipstick analysis of the clean-catch urine?

2. What would you report to the attending physician after the Dipstick of the "infected" urine?

3. Do the two dilutions give the same number of organisms/mL for each urine sample? Why might these numbers be different?

4. Does the number of colonies of bacteria per mL of the infected urine indicate a urinary tract infection? Do the results of the Dipstick test correlate positively with the culture result?

5. Various enteric bacteria, especially *E. coli*, are the cause of most urinary tract infections (UTI) in humans. A quick test for the presence of these bacteria in urine is to detect nitrite in the urine. Why?

6. Significant bacteria in a clean-catch urine specimen is >$10^5$/mL. Lower numbers are considered significant in urine collected by catheter or by suprapubic aspiration of the bladder. Why?

# EXERCISE 21

## Antiseptics and Disinfectants

## OBJECTIVES

At the conclusion of the exercise, you should...

1. know the difference between an antiseptic and a disinfectant.
2. be able to evaluate the bacteriostatic and bactericidal properties of bacteria.
3. be able to evaluate the relative effectiveness of various chemical substances that are advertised as antimicrobial agents.

## INTRODUCTION

There are a wide variety of chemical agents available that are supposed to control microbial growth. Two examples of antimicrobial agents are disinfectants and antiseptics. In this exercise, you will test the effectiveness of a variety of antiseptics and disinfectants on a Gram-positive bacterium and a Gram-negative bacterium.

## MATERIALS

Safety: Biosafety Level 1

Cultures:

*E. coli* - BSL1
*S. epidermidis* - BSL1

Media:

TSA plate

Supplies:

Sterile swabs
Sterile paper disks (¼-inch )
Pasteur pipettes
Miscellaneous antiseptics and disinfectants

## PROCEDURES
### Day 1 (Inoculation)

*Technical Background*

**Disinfectants** are agents (heat, radiation, or chemical) that inhibit the growth of disease-carrying microorganisms on inanimate objects. **Antiseptics** are substances that inhibit the growth and reproduction of disease-carrying organisms on living tissue.

When a disinfectant or antiseptic kills the microbe it comes in contact with, it is called a **bactericidal agent**. If it causes temporary inhibition of growth of the microbe, it is called a **bacteriostatic agent**.

*Preparing Inoculations*

1. Dip a sterile swab into the culture of *E. coli*.
2. Swab a TSA plate over the entire surface by going in one direction; then turn the plate and swab the entire plate going in the other direction.
3. Repeat swabbing the entire plate, swabbing in three directions (by turning the plate).
4. Choose two agents from the basket of miscellaneous antiseptics and disinfectants, or your own from home.
5. Using flamed forceps, remove a sterile ¼-inch disk from the vial and place it on the inside of the cover of the TSA plate. (See diagram.)
6. Use a Pasteur pipette to add a drop of a liquid agent to the disk, or squeeze a sample out of the container onto the disk.
7. Use the flamed forceps to transfer the saturated disk onto one-half of the TSA plate. Gently tap the disk so that it adheres to the agar.
8. Repeat the same procedure with another agent, and place the disk on the other half of the TSA plate.
9. Other students will follow the same procedure above with *Staphylococcus epidermidis*.
10. Invert all the plates and incubate at 35°C until the next lab period.

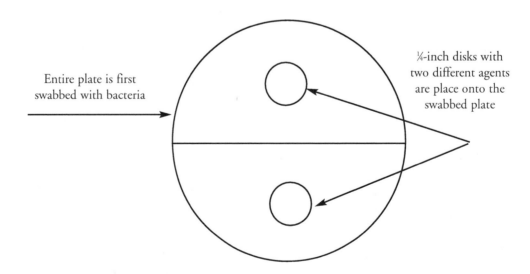

Entire plate is first swabbed with bacteria

¼-inch disks with two different agents are place onto the swabbed plate

## PROCEDURES
### Day 2 (Measurement of Disks)

*Evaluating Your Results*

1. Use a millimeter ruler to measure any zone around the disks on each of the plates. Measure diameter of disk.
2. Record the results in the Evaluation of Results section.
3. Compare your results with the other students in the class.

## EVALUATION OF RESULTS
## (EXERCISE 21: ANTISEPTICS AND DISINFECTANTS)

Purpose

Data

| Disinfectant/Antiseptic | *E. coli* Zone of inhibition in mm | *Staphylococcus epidermidis* Zone of inhibition in mm |
|---|---|---|
|  |  |  |
|  |  |  |
|  |  |  |
|  |  |  |

## CONCLUSIONS, DISCUSSIONS, AND QUESTIONS

1. What conclusions can be derived from your results? How do the results compare to the manufacturers' labels?

2. What factors will influence the zone size?

3. Were you able to determine whether the chemical was bacteristatic or bactericidal?

4. Devise an experiment to determine if a disinfectant or antiseptic is static or cidal?

5. Use your textbook to fill in the table below.

| Compound | Disinfectant or Antiseptic | Uses |
|---|---|---|
| Ethanol | | |
| Chlorine | | |
| Quaternary detergent | | |
| Phenol | | |

6.  Control of the growth of microorganisms may mean sterilization, disinfection (or sanitation), decontamination, or antisepsis (asepsis). Look up the definition for each of these terms in your lecture textbook and write the definitions in the spaces provided.

    Sterilization–

    Disinfection–

    Decontamination–

    Antisepsis–

# E X E R C I S E 22

## Antibiotic Sensitivity Testing

## OBJECTIVES

At the conclusion of the exercise, you should...

1. be able to use the disk diffusion method to determine antibiotic susceptibilities.
2. know what the minimum inhibitory concentration (MIC) represents.
3. become familiar with certain antibiotics and the groups of microorganisms they inhibit best.

## INTRODUCTION

To choose an appropriate antibiotic for the treatment of an infection, isolation of the microorganism is performed, followed by susceptibility tests. There are two different types of antibiotic susceptibility tests: broth dilution (micro and macro methods) and disk diffusion. The disk diffusion is also commonly known as the Kirby-Bauer method. In this exercise, you will learn a modified Kirby-Bauer method of disk diffusion antibiotic susceptibility testing.

## MATERIALS

Safety: Biosafety Level 1

Cultures:

*E. coli* - BSL1
*Pseudomonas stutzeri* - BSL1
*Salmonella LT2* - BSL1
*Staphylococcus epidermidis* - BSL1

Media:

Mueller Hinton agar plate:

Supplies:

Sterile swabs
Disk dispensers with selected antibiotic disks:

| Name of Antibiotic | Abbreviation on disk |
|---|---|
| Ampicillin | AM |
| Carbenicillin | CB |
| Cephalothin | CF |
| Erythromycin | E |
| Ciprofloxacin | CIP |
| Vancomycin | V |
| Tetracycline | Te |
| Tobramycin | NN |

# PROCEDURES
## Day 1 (Inoculation)

*Technical Background*

The broth dilution tests and the disk diffusion test are designed to determine the lowest concentration of an antibiotic that inhibits visible growth of a microbe in **vitro**, i.e., **The Minimum Inhibitory Concentration (MIC)**. The Kirby-Bauer method is a standardized system approved by the FDA and clinical laboratory standards committees.

A Petri plate containing Mueller Hinton agar for growth is inoculated uniformly over the entire surface of the plate. Paper disks that are impregnated with various antibiotics are placed on the surface of the agar. During incubation, the antibiotic diffuses into the agar from the disk from an area of high concentration to an area of low concentration. After incubation, the **effective** agent will inhibit bacterial growth and a **zone of inhibition** will form around the disk. The zone of inhibition is measured in millimeters and is compared to a standard table. The zone size determines if the antibiotic will be effective and can be correlated to an MIC value. The zone size can be affected by the concentration of the agar, the organism, the diffusion rate of the antibiotic, and the organism's growth rate. Mueller Hinton agar is an agar that allows the antibiotic to diffuse freely. The paper disks are impregnated with standardized concentrations by commercial companies.

*Preparing the Inoculations*

1. Label the MH plate with the organism that was assigned.
2. Aseptically swab the assigned culture onto the MH plate.
3. Swab the entire plate going in one direction; then turn the plate and swab the entire plate.
4. Repeat swabbing the entire plate, swabbing in three directions (by turning the plate).
5. Let the plate stand at least 5 minutes.
6. Use one of the disk dispensers and place a set of disks onto the plate by "pushing" the handle of the dispenser firmly over the agar.
7. Use a sterile loop or flamed forceps to touch each disk to ensure good contact with the agar.
8. Invert the plates and incubate at 35ºC until the next lab period.

## PROCEDURES
Day 2 (Measuring)

*Measuring the Zone of Inhibition*

1. Use a millimeter ruler to measure the diameter of the zone of inhibition.
2. Use the Antibiotic Zone Diameter interpretive table in the appendix to determine the significance of the zones of growth inhibition for the Kirby-Bauer Method of Antibiotic Sensitivity Testing.
3. Record your results in the Evaluation of Results section.

# EVALUATION OF RESULTS
# (EXERCISE 22: ANTIBIOTIC SENSITIVITY TESTING)

Purpose

Data

| Organism | | Ampicillin AM | Carbenicillin CB | Cephalothin CF | Erythromycin E | Ciprofloxacin CIP | Vancomycin V | Tetracycline TE | Tobramycin NN |
|---|---|---|---|---|---|---|---|---|---|
| E. coli | Zone | | | | | | | | |
| | Rating I, S, R | | | | | | | | |
| S. epidermidis | Zone | | | | | | | | |
| | Rating I, S, R | | | | | | | | |
| P. stutzeri | Zone | | | | | | | | |
| | Rating I, S, R | | | | | | | | |
| Sal LT2 | Zone | | | | | | | | |
| | Rating I, S, R | | | | | | | | |

## CONCLUSIONS, DISCUSSIONS, AND QUESTIONS

1. Which antibiotics are suitable for control of an infection with the four organisms that were tested?

2. Discuss the difference between narrow-spectrum and broad-spectrum antibiotics.

3. What factors will influence the size of the zone of inhibition?

4. Define MIC (minimum inhibitory concentration).

5. Indicate what types of bacteria are most effectively destroyed by each of the antibiotics tested.

# Isolation of Antibiotic-Resistant Mutants

## OBJECTIVES

At the conclusion of the exercise, you should…

1. understand that spontaneous mutations to antibiotic resistance occur in bacteria in the absence of the antibiotic.
2. know how to demonstrate antibiotic resistance by spontaneous mutation.

## INTRODUCTION

A **mutation** is any change in the nucleotide sequence of a gene. Most mutations are harmful because the gene product no longer works properly and the mutant cells quickly die off. Some mutations are neutral and neither harm nor enhance the functioning of the cell. However, when the cell finds itself in a new environment, e.g., in the presence of an antibiotic, then this "neutral" mutation, which occurred by chance, might – if the cell is lucky – allow the cell to survive and grow.

Consider a patient suffering from a bacterial infection. The doctor prescribes ampicillin, an antibiotic that inhibits the synthesis of the peptidoglycan in the bacterial cell wall. As the sensitive infecting bacteria grow, their walls weaken and they lyse. The body's natural defenses can now overwhelm the bacteria and clear them from the body: infection cured. However, by chance within the large population of bacteria, there may be a mutant cell that is not affected by the presence of circulating ampicillin. Chances are that the body will eliminate this mutant along with the sensitive bacteria. But if the patient stops taking the antibiotic early, this mutant may survive and grow. Then, when antibiotic therapy is resumed, the mutant can grow. The presence of circulating ampicillin will **select for** this mutant. This is an example of natural selection leading to the evolution of ampicillin resistance in this particular bacterium. Unfortunately for humans, once the mutant appears, it is often able to transmit the resistance gene to other bacteria.

In this exercise, you will use a non-pathogenic bacterium, *Serratia marcescens,* to prove that ampicillin-resistant cells appear by chance in the population <u>before</u> exposure to the antibiotic.

## MATERIALS

Safety: Biosafety Level 1

Culture:

A 24-30 hr. Luria broth (LB) 30°C culture of *Serratia marcescens* (or *E. coli*) wild type, diluted to about $1.2 \times 10^8$ cells/mL in phosphate-buffered saline and kept on ice. **Note: The bacteria in the broth culture have never been exposed to ampicillin.**

Media:

Plate of LB agar
Plate of LB agar + 100 micrograms/mL of ampicillin

Supplies:

L-shaped glass rod
Alcohol jar for flaming
Turntable
Canister of sterile 1-mL pipettes

## PROCEDURES

Day 1 (Inoculation)

Luria broth (LB) is a rich nutrient medium that supports the rapid growth of *Serratia marcescens*. Within a few hours, the inoculum of a few thousand cells will increase to about $5 \times 10^8$ cells per mL. Many of the cells will have spontaneous neutral mutations to any number of different genes, possibly ampicillin resistance among them. In the cozy, non-selective environment of LB, the cells with neutral mutations persist and grow. Once the population is placed onto ampicillin-containing agar plates, the sensitive cells will die, and only the few resistant mutants already present will survive and form colonies.

*Preparing the Inoculations*

1. Label the LB plate and the LB + ampicillin plate appropriately.
2. Pipette 0.1 mL of the culture onto the surface of each plate.
3. Take the L-shaped glass rod out of the alcohol, and set it alight.
4. Wait 20 seconds or so for the rod to cool.
5. Place one plate on the turntable and spread out the liquid evenly.
6. Repeat with the other plate.
7. Return the glass rod to the alcohol.
8. Leave the plates right side up on the benchtop for about 10 minutes.
9. Incubate upside down at 35°C until the next period.

## PROCEDURES
## Day 2 (Observations)

Recall that usually a colony forms after many rounds of growth and division of the single cell that was originally deposited on the agar surface. Also, recall that ampicillin interferes with the synthesis of the cell wall of bacteria. Therefore, ampicillin-sensitive bacteria on the LB + ampicillin plates cannot make cell walls, and they lyse. No colonies form.

### Counting Colonies

1. Count the number of colonies on both the LB control plate and the LB + ampicillin plate.
2. Record your data in the Evaluation of Results section.
3. Estimate the rate at which ampicillin-resistant spontaneous mutants occur in a population of *Serratia marcescens*.

   Example of calculation: 10 streptomycin$^R$ colonies appear on an LB + streptomycin plate that was inoculated with $1 \times 10^7$ bacteria. What is the spontaneous mutation rate to Str$^R$?

   10 Str$^R$ mutants in a population of $1 \times 10^7$ bacteria, or $10 \div 1 \times 10^7 = 1 \times 10^{-6}$
   (In other words, 1 organism out of 1 million is the spontaneous mutation rate.)

## EVALUATION OF RESULTS
## (EXERCISE 23: ISOLATION OF ANTIBIOTIC-RESISTANT MUTANTS)

Purpose

Data

### Results of Selection for Antibiotic-Resistant Mutants

| Number of cells plated | Number of colonies on LB agar | Number of colonies on LB + ampicillin agar |
|---|---|---|
|  |  |  |

Use TNTC for "too numerous to count."

## CONCLUSIONS, DISCUSSION, AND QUESTIONS

1. What is the rate of spontaneous mutation to ampicillin resistance in this bacterium? Show calculation.

2. Which agar plate was the negative control?

3. Why was this plate included in the experiment?

4. Sneedy is a non-compliant patient. He takes the antibiotic for his bacterial infection for only 3 of the 8 days prescribed by the doctor. This is not long enough for his body to completely eradicate all of the infecting bacteria from his system, and they begin to grow again. Then he starts taking the antibiotic again (because he is feeling sick). The antibiotic is no longer effective, and his doctor has to switch him to a different antibiotic (after a stern lecture on the importance of finishing the course of treatment). **Explain** why the antibiotic was not effective the second time around.

5. Unfortunately for humanity, Sneedy's antibiotic-resistant bacteria carry the gene for resistance on a small, mobile piece of DNA called a plasmid. Plasmids are able to move from one bacterium to another, as long as the bacteria are related to each other. Sneedy ends up in the hospital for intravenous antibiotic treatment. His nurse forgets to wash his hands after caring for Sneedy and moves the resistant bacteria to another patient. **Speculate** as to what might happen next.

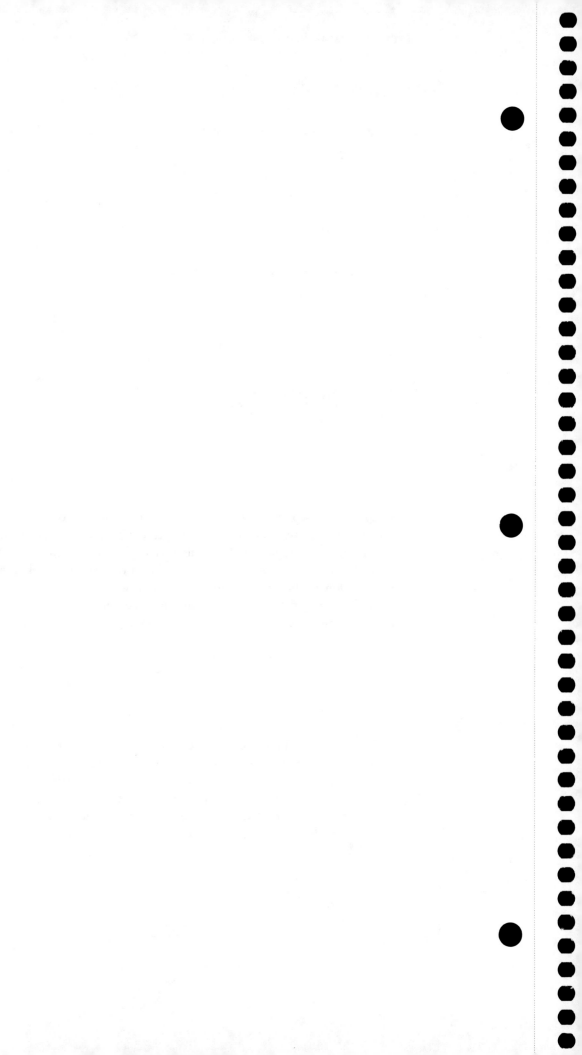

# Unknowns

## OBJECTIVES

At the conclusion of the exercise, you should...

1. be able to use the techniques learned so far and apply critical thinking for identifying an unknown sample.
2. understand the relevance of identifying unknown organisms from clinical or environmental samples.
3. learn how to keep stock and working cultures of bacteria in a culture collection for testing.
4. understand how to maintain a pure culture and how to recognize a contaminant.
5. learn how to write a lab report using scientific format.

## INTRODUCTION

This exercise provides students with the opportunity to apply all of the previous laboratory techniques and bacteria studied so far for identifying unknown bacteria. It can be one of the most interesting experiences in microbiology. You will be given cultures of bacteria to identify and some information about the source of the culture. The identification of the unknown will be performed through independent study, by applying all the techniques learned thus far. The first unknown is taught as a class so students have some practice working with unknowns. Each student will write his or her own Unknown lab report. The second unknown is the real unknown, and each student works independently.

## MATERIALS

Safety: Biosafety Level 1

Media

BCP Lactose
TSA plates and slants
Tryptone broth
Citrate
Methyl red broth
PIA

## PROCEDURES
## UNKNOWN #1 (DONE AS A CLASS FOR PRACTICE)
### Day 1 (Inoculations)

*Technical Background*

You will be provided four different cultures of bacteria, labeled A through D. The specimen source or body site will be included. There will be 3 of each for a total of 12. This will be done in pairs. One person from each pair will choose one of the tubes and record which letter they took and the specimen source. Each student will streak a TSA plate with the one unknown. Follow the directions below and use the worksheet. This will be done as a class exercise. Remember to use the unknown worksheets provided at the end of this exercise and the identification tables in the appendix. Check with the instructor for the appropriate media. **SAVE** everything until you are completely finished with your practice unknown. This includes the original cultures and all stained slides, media test tubes, and plates. Remember that all test tube media should be inoculated with loose caps(!) and incubate plates upside down! And you should be labeling all tubes and plates properly.

*Growing a Culture of the Practice Unknown*
(done singly)

1. Streak the unknown onto a TSA plate (streak for isolation) and incubate at 35ºC (48 hrs.).
2. Perform a Gram stain from the broth, if desired (time permitting). Remember to do this on a QC slide.
3. Save the original unknown TSB tubes in the refrigerator.

## PROCEDURES
### Day 2

*Preparing Inoculations of the Practice Unknown*

1. Perform a Gram stain on a fresh isolated colony.
2. Determine which unknown identification table to use.
3. Pick the rest of the colony and two or three others that appear the same and inoculate one TSA slant (this will be the working slant).
4. Incubate the TSA slant and the BCP lactose at 35ºC until the next lab period.
5. Store the original broths and the TSA plates in the refrigerator.

## PROCEDURES
Day 3

*Interpreting the Lactose Reaction of the Practice Unknown*

1. Read and record the BCP lactose broth reaction.
2. **If the organism is Lactose Positive:** Then inoculate from the TSA slant into Tryptone broth (for Indole), Citrate, Methyl Red (MR).

3. **If the result is Lactose Negative:** Then inoculate from the TSA slant into Tryptone broth, Citrate, Methyl Red (MR), and PIA (Peptone Iron agar). Note: Because the oxidase test should be done on a fresh culture, it was already done on Day 2. Make sure that this was recorded correctly on the worksheet.

## PROCEDURES
Day 4
Each student reads, records, and writes his or her own lab report based on all of the results.

*Recording and Evaluating Results on the Practice Unknown*

1. Read and record all of the tests. Record everything on the worksheet. Use the identification tables in the Appendix to determine the identification of your practice unknown.
2. Give your results to the TA.
3. Discuss the results as a class.

## PROCEDURES
Day 5

*Writing Your Lab Report on Practice Unknown*

1. Finish any tests not completed earlier.
2. Read the appendix on writing in scientific format.
3. Class discussion about writing a lab report in scientific format.
4. Turn in Lab Report #1 on the due date.

**The Evaluation of results for the unknown exercise** will be the written Unknown Report #1. Each student writes his or her own practice unknown lab report, using the results obtained as a class. Follow the directions of the instructor and the information in the Appendix for writing lab reports using the scientific format. Use the Unknown #1 worksheets (two: one rough, one final) at the end of the exercise. Turn in both worksheets with your written report, along with flow charts. Remember to include information about the organism, the specimen source, and its relationship in disease.

## UNKNOWN #1 WORKSHEET
### (Include this worksheet and the flow charts with your written report)

Name: _____

Lab section: _____

Date specimen received: _____

Specimen number: _____

Gram reaction and morphology:

_____

_____

_____

Colonial morphology:

_____

_____

_____

Specimen source: _____

Final identification: _____
Use the back of this sheet or a separate sheet to make a flow chart of your tests and results.

# UNKNOWN #1 WORKSHEET (FINAL)
### (Include this worksheet and the flow charts with your written report)

Name: _____

Lab section: _____

Date specimen received: _____

Specimen number: _____

Gram reaction and morphology:

_____

_____

_____

Colonial morphology:

_____

_____

_____

| Differential Tests and Results: ||
|---|---|
| **TEST** | **RESULT** |
|  |  |
|  |  |
|  |  |
|  |  |
|  |  |
|  |  |
|  |  |
|  |  |
|  |  |
|  |  |
|  |  |
|  |  |
|  |  |
|  |  |

Specimen source: _____

Final identification: _____
Use the back of this sheet or a separate sheet to make a flow chart of your tests and results.

## PROCEDURES
## UNKNOWN #2
### Day 1 (Inoculations):

*Technical Background*

Unknown #2 is done individually. Remember, this is a critical thinking exercise and the objective is for you to be able to think and solve a problem using all the information that has been taught to you thus far. An unknown lab report, entitled Unknown #2, will be written, following scientific format.

*Identifying Unknown #2*

1. Follow the instructions of the TA for choosing unknown #2.
2. Sign on the blank sign-up sheet.
3. Record your unknown number and the specimen source listed.
4. Streak each unknown onto a TSA plate (streak for isolation) and incubate for 24-48 hrs. at 30°C.
5. Place all plates on a tray for your section so that they can be moved to the refrigerator after 48 hours, if necessary.
6. Save the original cultures in the refrigerator, and save all the plates and biochemical reactions in test tubes, until the end of the exercise.
7. Use the Unknown worksheets in the back of this exercise to record your results. Make sure you record your unknown numbers.

## PROCEDURES
Day 2  (Gram Stains from TSA Plates and Inoculation of Working Slants)


Determine the Gram reaction and shape of each of your unknowns from a pure isolated colony on the TSA plate. Inoculate a working TSA slant to use for inoculating the biochemical tests necessary for identification. **Also inoculate a BCP Lactose broth.** After the Gram reaction is determined, choose the appropriate identification table to use for the identification of the unknown. All the tests need to be done from the working TSA slant after it has grown for 24-48 hours. Note: It is **not** necessary to perform all the tests listed in the identification tables. You will be given a list of media available for Gram-negative and Gram-positive bacteria.

Reminders:     Label everything appropriately
Incubate with loose caps
Save all tubes and plates in the refrigerator
Save all stained slides

NOTE: In order to get any media, you will always have to write down the media that you are requesting on each of your worksheets and get your TA's initials. Media needed outside of lab class time must have a TA's note or initials on the worksheet before being requested from the prep room. Please remind your TA of this before you leave the classroom.

**Reminder: Save everything! This includes the Gram-stained slides and all test tubes and plates after reading, recording, transferring, inoculating, etc.**

## PROCEDURES
### Day 3  (Confirmation of Gram Reaction and Purity Check)

Inoculate the appropriate media to determine the identification of the unknown bacteria. Remember to go back to your notes for each of the tests and review all the tests learned thus far.

**Available Media and Supplies:**
> TSB broths
> TSA plates
> Gram stain kits
> Hydrogen peroxide (3%)
> Kovac's reagent
> Methyl Red reagent
> Nitrate test reagents
> Oxidase paper
> 1 set uninoculated unknown media

**Gram-Negative Bacteria:**
> BCP Adonitol
> BCP Lactose
> BCP Mannitol
> BCP Raffinose
> BCP Sucrose
> Tryptone broth (Indole test)
> Citrate
> MR (Methyl Red) broth
> PIA (Peptone Iron Agar) stab for $H_2S$

**Gram-Positive Bacteria:**
> BCP Lactose
> BCP Mannitol
> BCP Raffinose
> BCP Sucrose
> Citrate
> Hydrogen peroxide
> MR (Methyl Red) broth
> Nitrate

**Special media available only if there is difficulty in determining the identification:**
> Urea slant
> Motility stab
> Coagulase plasma
> DNAse plate

## PROCEDURES
### Day 4 (Results)

*Evaluating Your Results and Writing Your Report*

1. Read and record all the test results. Remember to go back to your notes for each of the tests and review what reagents, if any, are necessary, and what a positive and negative reaction looks like.
2. Using the identification tables provided, determine the identity of your unknown organism.
3. Remember to record everything on the worksheets, and **save** everything until you are completely finished.
4. The Evaluation of Results for this exercise will be the written Unknown Report #2. Follow the directions of the instructor and the information in the Appendix for writing lab reports using the scientific format. The due date for the written report is listed in the syllabus.
5. Use the Unknown #2 worksheets (2 provided: 1 rough, 1 final).
6. Use flow charts and tables in the final report. (See examples.)
7. Turn in all of the worksheets with the written report.

# UNKNOWN #2 WORKSHEET
### (Include this worksheet and the flow charts with your written report)

Name: _____

Lab section: _____

Date specimen received: _____

Specimen number: _____

Gram reaction and morphology:

_____

_____

_____

Colonial morphology:

_____

_____

_____

| Differential Tests and Results: | |
|---|---|
| **TEST** | **RESULT** |
|  |  |
|  |  |
|  |  |
|  |  |
|  |  |
|  |  |
|  |  |
|  |  |
|  |  |
|  |  |
|  |  |
|  |  |
|  |  |

Specimen source: _____

Final identification: _____
**Use the back of this sheet or a separate sheet to make a flow chart of your tests and results.**

# UNKNOWN # 2 WORKSHEET (FINAL)
### (Include this worksheet and the flow charts with your written report)

Name: _____

Lab section: _____

Date specimen received: _____

Specimen number: _____

Gram reaction and morphology:

_____

_____

_____

Colonial morphology:

_____

_____

_____

| Differential Tests and Results: | |
|---|---|
| **TEST** | **RESULT** |
| | |
| | |
| | |
| | |
| | |
| | |
| | |
| | |
| | |
| | |
| | |
| | |
| | |

Specimen source: _____

Final identification: _____
Use the back of this sheet or a separate sheet to make a flow chart of your tests and results.

# EXERCISE 25

## Immunology

## OBJECTIVES

At the conclusion of the exercise, you should...

1. be able to define the terms antigen and antibody.
2. be able to define agglutination and titer.
3. know the purpose for doing agglutination tests in immunology.
4. be able to read and interpret a tube agglutination reaction.
5. be able to perform a slide agglutination test.
6. understand how latex agglutination tests work.
7. understand hemagglutination.
8. be able to identify and discuss the cell types found in human blood.

## INTRODUCTION

**Antigens** are substances (usually proteins) that are capable of stimulating **antibody** production when injected into an animal. An **antibody** is a **specific** substance that is produced in the blood or tissues in response to a **specific** antigen. It functions to overcome the effect of the antigen, such as toxins (antigens). In this exercise, you will observe antigen-antibody agglutination reactions and learn their significance in immunology.

## A. SLIDE AGGLUTINATION

## MATERIALS

Safety: Biosafety Level 1

Cultures:

*Salmonella* species antigens (killed and preserved suspensions) - BSL1
*E. coli* or some other enteric suspension (antigens) - BSL1
Unknown antigens (optional)

Supplies:

*Salmonella* species antisera
Sterile applicator sticks

0-20 µl micro pipettes
micro pipette tips
micro pipette waste beaker

## PROCEDURES
### Day 1

*Technical Background*

Blood serum is the clear yellow portion of blood after it has clotted and separated from the red blood cells. **Antiserum** or **antisera** (plural) contains **antibodies**. It is obtained from animals that have been injected with antigenic materials such as toxins or bacteria to stimulate the production of antibodies. Bacterial cells possess antigens on their cell surfaces. If a bacterial cell invades the body, antibodies are produced specifically against that organism. If another microbe enters the body, new specific antibodies are produced against the second invader. The first set of antibodies will not react with the second microbe, nor will the second antibodies react with the first antibodies produced. Antigen–antibody reactions are very specific. So if a drop of *Salmonella typhimurium* antiserum is mixed with a drop of an unknown organism and agglutination occurs, then the identification of the unknown organism has been confirmed, because only *S. typhimurium* will agglutinate with *S. typhimurium* antibodies (antiserum). There is typing sera available for many species of bacteria, and the reaction is used to aid in identifying mostly Gram-negative enteric pathogens.

*Slide Agglutination*

1.  Label a microscope slide, as shown in the diagram below.
2.  Remove 20 µl aliquots of antisera and place in two separate spots on the clean slide.
    **Note: Always use a new sterile tip every time you remove a sample from the antisera and antigen.**
3.  Using a new sterile tip, add 20 µl of the Salmonella antigen to the first drop of antiserum (Positive control).
4.  Using a new sterile tip, add 20 µl of the *E. coli* antigen to the second drop of antiserum (Negative control).
5.  Mix each area with a clean applicator stick or toothpick.
6.  Gently rock the slide back and forth and in a circular pattern, being careful not to allow the areas to mix together. Mix up to two minutes.
7.  Observe for agglutination in each spot.
8.  Record the results in the Evaluation of Results section.
9.  If unknown antigens are available, perform a slide agglutination on a separate slide (drop of unknown antigen plus drop of Salmonella antiserum).
10. Answer the questions in the Evaluation of Results section.

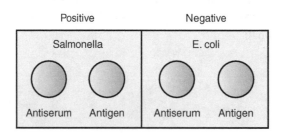

## B. TUBE AGGLUTINATION

## MATERIALS

Safety: Biosafety Level 1

Cultures:

Demonstration:

Serial tube agglutination test

## PROCEDURES
Day 1

*Technical Background*

**Agglutination titration** can estimate the concentration of antibodies in serum to a specific anti-gen. The patient's serum is diluted in saline in a series of tubes, starting with a 1:10 dilution. A twofold **serial dilution** is usually done. (See diagram). The suspected antigen is added to all the tubes. The tubes are mixed and incubated at 35°C for 30-60 minutes. Each of the tubes is examined for agglutination. The tube with the highest dilution at which agglutination occurs is the **antibody titer**. The antibody titer of the patient's serum is expressed as the **reciprocal of the highest dilution**. For example, if the tube with the 1:320 dilution is the last tube with agglutination, then the **antibody titer** is 320. The patient's serum has 320 antibody units per millimeter of serum for the disease being tested.

This serial dilution tube test was first done with serum from patients with typhoid fever (*Salmonella typhi*) by Grunbaum and Widal, in the 1800s. Thus, it is referred to as the **Widal Test**. The technique has been adapted for many other diseases.

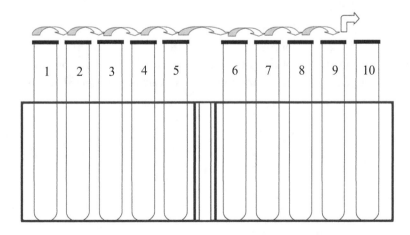

0.5 mL transferred from tube to tube                    0.5 mL discarded from tube 9

1:10  1:20   1:40  1:80   1:160        1:320 1:640 1:1280 1:2560  Control
0.5 mL saline per tube

Start with 1 mL of a 1:10 dilution of the patient's serum. Then, 0.5 mL is transferred from tube to tube, with the last 0.5 mL being discarded. Last, 0.5 mL of the antigen is added to each tube for the final dilutions.

*Performing a Widal Test*

1. Observe the Tube Agglutination Test in the laboratory.
2. Gently shake each of the test tubes to see any agglutination.
3. Record the antibody titer in the Evaluation of Results section.

# C. LATEX SLIDE AGGLUTINATION

## MATERIALS

Safety: Biosafety Level 1

Supplies:

If available-commercial latex agglutination test kits for:
      Streptococcus

## PROCEDURES
Day 1

*Technical Background*

**Streptex** (Wellcome Diagnostics): The majority of species of *Streptococcus* possess group-specific antigens that are usually carbohydrate structural components of the cell wall. These antigens can be extracted in soluble form and identified by precipitation reactions with antisera.

The latex suspension used in this kit is made of polystyrene latex particles, which are coated with purified rabbit antibody to the appropriate group antigen. The extraction enzyme is used to extract the antigens from the cell wall of the streptococci. The positive control is a polyvalent extract containing antigens from a representative strain of each streptococcal group A, B, C, D, F, and G.

# D. BLOOD GROUPING

## MATERIALS

Supplies:

Unknown simulated blood types
Blood typing sera (simulated): Anti-A, Anti-B, Anti-D (Rh)
Shields
Gloves
Applicator sticks or toothpicks

## PROCEDURES
Day 1

*Technical Background*

Another type of agglutination reaction can occur with the antigens on red blood cells and the antibodies in human serum. This type of reaction is called **hemagglutination**. Hemagglutination reactions are used in typing blood. Four major blood types have been identified. Type A individuals possess A antigens on their red blood cells. Type B have B antigens. Type AB have both A and B antigens, and Type O individuals have neither A nor B antigens on their red blood cells. The Rh factor is another antigen found on the red blood cells in about 85% of the population. The Rh factor is a complex of many antigens, one of which is D. The D antigen is what is tested for to determine the Rh factor of an individual.

A person possesses antibodies to the opposite antigen found on their red blood cells. So, people with type A red blood cell antigens will have antibodies to the B antigen in their sera. Type B will have antibodies to A antigens, and Type AB will not have A or B antibodies in their serum. Type O individuals will have both anti-A and anti-B antibodies in their serum. The hemagglutination reaction is important in blood banking, prior to using any blood product for transfusions.

**All work is to be done behind the plexiglass shield:**
1. Draw two circles on a clean microscope slide.
2. Draw one circle on a second microscope slide.
3. Squeeze a drop of the assigned unknown blood onto the slide into each of the three circles.
4. Add one drop of Anti-A next to one drop of blood on the first slide.
5. Add one drop of Anti-B next to the other drop of blood on the first slide.
6. Add one drop of Anti-D next to the other drop of blood on the second slide.
7. Mix the drops with toothpicks or applicator sticks.
8. Place the second slide on the slide warmer and gently rock back and forth and watch for agglutination.
9. Rock the first slide and watch for agglutination.
10. Discard all materials in the SHARPS container.
11. Determine the blood type using the table.

| Tested Blood (antigens) | Anti serum | | | Blood Type and Rh | Percent Occurrences* |
|---|---|---|---|---|---|
| | Anti-A | Anti-B | Anti-D | | |
| A | + | - | + | A positive | 35% |
| | + | - | - | A negative | 6% |
| B | - | + | + | B positive | 8.5% |
| | - | + | - | B negative | 1.5% |
| AB | + | + | + | AB positive | 3.4% |
| | + | + | - | AB negative | < 1% |
| O | - | - | + | O positive | 38% |
| | - | - | - | O negative | 6.8% |

* Based on 41% are Type A, 10%; Type B, 4%; Type AB, 45%; Type O, 85% are Rh (D) positive; 15% are Rh (D) negative.

## E. WHITE BLOOD CELLS

## MATERIALS

Safety: Biosafety Level 1

Supplies:
Normal peripheral blood smears – Wright stained

## PROCEDURES
Day 1

*Technical Background*

There are three types of cellular elements present in blood: red blood cells (erythrocytes), white blood cells (leukocytes), and platelets (thrombocytes). Each of these cells has its own function, and are morphologically different.

The normal adult has about 6 L of blood, which composes 7-8% of the total body weight. Plasma (the liquid portion of blood), makes about 54% of blood volume, 45% is composed of erythrocytes, and 1% is leukocytes and thrombocytes. Variations in blood elements are often the first sign of disease occurring in body tissue.

### Normal adult blood values:

Red blood cells: 3.8–5.9 x $10^9$/l
Platelets: 150–440 x $10^9$/l
White blood cells: 3.5–11.0 x $10^9$/l

### White blood cell differential count:

The differential white blood cell count is performed to determine the relative number of each type of white blood cell present in the blood. At the same time, a study of red blood cell, white blood cell, and platelet morphology is performed. More information can be obtained from a detailed examination of the stained blood smear than from any other single laboratory test.

### Types of white blood cells and normal adult ranges:

|  |  |
|---|---|
| neutrophils: | 50–70% |
| eosinophils: | 0–4% |
| basophils: | 0–2% |
| monocytes: | 2–9% |
| lymphocytes: | 20–44% |

**Neutrophil:**
light-pink granules in cytoplasm with a purple nucleus with segments and lobes
size: 10–16 mm
major function of ingesting and killing microorganisms.

**Eosinophil:**
large, pink granules fill the cytoplasm and surround the purple segmented, lobed nucleu
size: 10–16 mm
one function is to provide some defense against parasitic infections

**Basophil:**
large, purple-black granules, sometimes covering the dark purple nucleus
size: 10–16 mm
one function is to participate in allergic reactions

**Monocyte:**
gray cytoplasm with vacuoles (holes), with a fine, large, purple nucleus
size: 12–20 mm.
major function is ingestion of dead or dying cells

**Lymphocyte:**
sky-blue cytoplasm (usually very little), with a large, purple nucleus
size: 7–18 mm
vital to the immune system, by producing antibodies and providing cellular immunity

**Red blood cells (erythrocytes):**
pink-stained cells; some may show clear center
size: 6–8 mm
primary function is to carry hemoglobin, which, in turn, transports oxygen to the tissues and carbon dioxide from the tissues to the lungs

**Platelets (thrombocytes):**
are not complete cells; they are fragments from a cell in the bone marrow called a megakaryocyte
very small, irregular-shaped, purple-stained particle
size: 2–4 mm
function primarily in hemostasis - (the stoppage of bleeding)

*Identifying Blood Cells*

1. Observe the different blood cells on the stained slides.
2. Make sure you can tell the difference between a white blood cell, a platelet, and a red blood cell.

# EVALUATION OF RESULTS
# (EXERCISE 25: IMMUNOLOGY)

Purpose

Data

A.  Slide Agglutination

Control                                               Organism

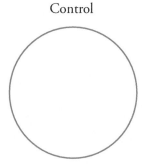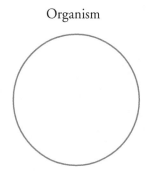

B.  Tube Agglutination

Titer =

C.  Latex Agglutination (if available)

Description of your observations:

D.  Blood Typing

Unknown Blood Type:

E.  White Blood Cell Differential

Description of your observations:

## CONCLUSIONS, DISCUSSIONS, AND QUESTIONS

1.  Define Titer–

2.  Define Antibody–

3.  Define Antigen–

4.  You should have seen agglutination on the slide with the Salmonella antigen and the Salmonella antiserum and not with the *E. coli* antigen and Salmonella antiserum. Why?

5.  If you had performed the slide agglutination with an unknown antigen that did not show agglutination, but the positive control did and the negative control did not, what could you deduce about the unknown? If the unknown did show agglutination, like the positive control, but the negative control did not, what can you deduce about the unknown bacteria?

6.  What was the titer of the tube agglutination demonstration?

7.  An agglutination test was performed to find the concentration of antibody in a patient's serum. Antigen attached to particles was added to all wells. Dilutions of the patient's serum was added to all but the control well.

Agglutination +

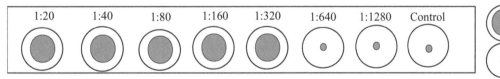

Agglutination –

Define Serum–

Has the patient shown seroconversion (i.e., production of antibodies to the antigen)?

What is the antibody titer?

One year after the patient fully recovered, another agglutination test was performed on his serum. Using the same antigen as before, the following results were obtained:

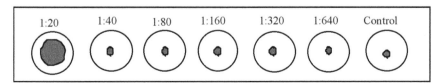

What is the antibody titer?

What is your interpretation of this result?

8.  Why are people with Blood Group AB called "universal recipients"?

# Environmental and Food Microbiology

This section introduces the techniques used in water and food microbiology. Again, the techniques learned from the previous exercises will be applied, plus some new ones. The first exercise is the standard technique used for examining water for contamination. It is used for drinking water and for recreational waterways. When you hear on the news that a beach has closed due to sewage spills, the technique that is used to determine this is the MPN procedure.

Microbes can be responsible for spoiling foods and causing diseases. The spoilage of meat exercise and the standard plate count of milk have to do with microbial contamination and techniques used to determine the spoilage.

Microorganisms have been used for making foods for many years. Two examples that you will perform are wine-making and yogurt production. You will even get the chance to taste the food that you make.

Safety: The laboratory exercises in this section will be using Biosafety Level 1 and Biosafety Level 2 procedures.

# EXERCISE 26

## Water Microbiology

## OBJECTIVES

At the conclusion of the exercise, you should...

1. be able to list three techniques to determine water quality.
2. understand how water quality is determined using indicator bacteria.
3. know what a coliform is.
4. be able to perform the multiple tube technique for determining water quality.

## INTRODUCTION

Infectious diseases can be transmitted by contaminated water. Some of these diseases include typhoid fever, cholera, traveler's diarrhea, protozoan diseases, and some viral diseases. It is very difficult to test water for the causative agents of all these diseases. Tests have been developed to determine the bacteriologic quality of water using **indicator organisms** to detect fecal contamination of water. The most common indicator organisms used are the **coliform bacteria**. In this exercise, you will learn three techniques for detecting fecal coliform contamination in water: the Most Probable Number Method (MPN), the Millipore Filter Technique, and the MUG test.

## A. MPN (MOST PROBABLE NUMBER)

## MATERIALS

Safety: Biosafety Levels 1 and 2

Cultures:

Water sample labeled A
Water sample labeled B
Water sample labeled C

Media:

2X (double-strength) lactose broths with Durham tubes
1X (single-strength) lactose broths with Durham tubes
EMB plate
TSA slant
1X lactose broth

Supplies:

Pipettors
1-mL Sterile pipettes
10-mL Sterile pipettes

## PROCEDURES
### Day 1 (Inoculation)

*Technical Background*

**Coliforms** are Gram-negative, non-spore-forming rods that are facultative anaerobes and ferment lactose with acid and gas within 48 hours at 35°C. They are not usually pathogenic and are commonly found in the human gastrointestinal system. *E. coli* is the most well-known coliform. The Public Health Department has established standards for the maximum number of coliforms allowable in each 100 mL of water, depending on the intended use (such as drinking, water sports, or irrigation).

The bacterial examination of water using the **multiple tube technique** involves the following tests:

**Presumptive test:** In this test, a lactose-containing broth medium is inoculated with different amounts of the water sample. Three double-strength lactose broths receive 10 mL each. Three single-strength lactose broths receive 1 mL each. The remaining three single-strength lactose broths receive 0.1 mL each. All are incubated 24-48 hours at 37°C. A positive test is indicated by a greater-than 10% gas accumulation in the Durham tube. Because of the series of tubes used and the different sizes of inocula, the most probable number (MPN) of lactose-fermenting bacteria per 100 mL of water sample can be determined by referring to the probability table.

**Confirmed test:** False positives may occur in the presumptive test since some *Clostridium* and *Bacillus* species also ferment lactose, producing gas. False positives may also occur due to synergistic reactions of Gram + cocci with *Proteus* sp. To rule out any false-positive findings, you will inoculate an eosin methylene blue (EMB) plate by the normal streak method from the tube with the smallest inoculum and the most gas of that series. The eosin methylene blue plate will inhibit the growth of Gram-positive bacteria. *E. coli* will develop a dark color with a marked greenish sheen. *Enterobacter aerogenes* will form a large, mucoid, purple colony with whitish edges.

**Completed test:** Any suspected *E. coli* colony is then inoculated into a tube of lactose broth. It should produce acid and gas in this pure culture.

*Preparing the Inoculations*

1. Label a test tube rack or beaker with the water sample letter.
2. Label three single-strength (1X) lactose broth tubes with 1.0 mL.
3. Label three single-strength (1X) lactose broth tubes with 0.1 mL.
4. Label three double-strength (2X) lactose broth tubes with 10 mL.
5. Shake the water sample bottle and transfer 10 mL into each of the three 2X lactose broth tubes.
6. Transfer 1 mL of water into each of the three 1X lactose broth tubes.
7. Transfer 0.1 mL of water into each of the last three 1X lactose broth tubes.
8. Incubate the tubes at 35°C until the next lab period.

## PROCEDURES
Day 2 (Presumed Test Results and Confirmed Test Inoculations)

*Technical Background*

Most Probable Numbers (MPN) Table.
Adapted from the Standard Methods for the Examination of Water and Wastewater.

| Number of tubes giving positive reaction out of | | | MPN per 100 mL | Number of tubes giving positive reaction out of | | | MPN per 100 mL |
|---|---|---|---|---|---|---|---|
| 3 of 10 mL each | 3 of 1 mL each | 3 of 0.1 mL each | | 3 of 10 mL each | 3 of 1 mL each | 3 of 0.1 mL each | |
| 0 | 0 | 0 | <3 | 3 | 0 | 0 | 23 |
| 0 | 0 | 1 | 3 | 3 | 0 | 1 | 39 |
| 0 | 1 | 0 | 3 | 3 | 0 | 2 | 64 |
| 1 | 0 | 0 | 4 | 3 | 1 | 0 | 43 |
| 1 | 0 | 1 | 7 | 3 | 1 | 1 | 75 |
| 1 | 1 | 0 | 7 | 3 | 1 | 2 | 120 |
| 1 | 1 | 1 | 11 | 3 | 2 | 0 | 93 |
| 1 | 2 | 0 | 11 | 3 | 2 | 1 | 150 |
| 2 | 0 | 0 | 9 | 3 | 2 | 2 | 210 |
| 2 | 0 | 1 | 14 | 3 | 3 | 0 | 240 |
| 2 | 1 | 0 | 15 | 3 | 3 | 1 | 460 |
| 2 | 1 | 1 | 20 | 3 | 3 | 2 | 1,100 |
| 2 | 2 | 0 | 21 | 3 | 3 | 3 | >2,400 |
| 2 | 2 | 1 | 28 | | | | |

*Evaluating Your Results*

1. Shake the three 10-mL lactose broth tubes.
2. Repeat with the 1-mL lactose broth tubes and the 0.1-mL tubes.
3. Examine all 9 tubes for growth, acid production, and gas.
4. Record the results in the data section of Evaluation of Results.
5. Use the MPN table provided to determine the most probable number of coliforms per 100 mL of the water sample.
6. Then choose the lactose tube with the highest dilution and showing positive gas in the Durham tube.
7. Inoculate a loopful from the positive tube onto an EMB agar plate.
8. Streak for isolation.
9. Invert and incubate at 35°C until the next lab period.

## PROCEDURES
### Day 3 (Completed Test Results)

*Preparing the Inoculations*

1. Examine the EMB plates for coliform colonies.
2. Choose a typical coliform colony (see description and notes from previous demonstrations).
3. Inoculate a 1X lactose broth tube.
4. Incubate at 35°C until the next lab period.

## PROCEDURES
### Day 4 (Positive Test Confirmation)

*Evaluating Your Results*

1. Examine the lactose broth for gas formation.
2. Enter all results in the Evaluation of Results section.

## B. MILLIPORE FILTER

## MATERIALS

Safety: Biosafety Levels 1 and 2

Cultures:

Water (100-mL) sample bottle with *E. coli*

Media:

mEndo MF medium (2 mL)

Supplies:

Gelman filter unit
Vacuum flask
Vacuum pump
Trap
Hose
Sterile 5-mL pipette canister

# PROCEDURES
## Day 1

*Technical Background*

A method utilizing the membrane filter has been recognized by the U.S. Public Health Service as a reliable method for the detection of coliforms in water. These filter disks are 150-µm thick, have pores of 0.45-µm diameter, and have 80% area perforation. The precision of manufacture is such that bacteria larger than 0.47 µm cannot pass through. Eighty percent area perforation facilitates rapid filtration.

To test a sample of water, the water is passed through one of these filters. All bacteria present in the sample will be retained directly on the filter's surface. The membrane filter is then placed on an absorbant pad saturated with liquid nutrient medium and incubated for 22-24 hours. The organisms on the filter disk will form colonies that can be counted under the microscope. If a differential medium, such as mEndo MF broth is used, coliforms will exhibit a characteristic metallic sheen.

The advantages of this method over the multiple tube test are 1) higher degree of reproducibility of results; 2) greater sensitivity, since larger volumes of water can be used; and (3) shorter time (one-fourth) for getting results.

*Performing the Millipore Filter Method Test*

1. The Millipore Filter Method will be demonstrated.
2. After incubation at 35ºC, observe and count the colonies on the plate.
3. Enter your results in the Evaluation of Results section and answer the questions.

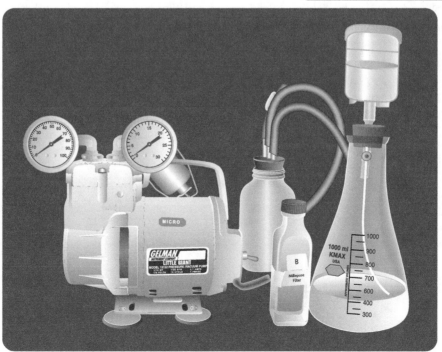

## C. MUG METHOD

## MATERIALS

Safety: Biosafety Levels 1 and 2

Cultures:

Various bacteria inoculated in the MUG media

Media:

MUG broth

Supplies:

UV light source

## PROCEDURES

Day 1

*Technical Background*

A method has been developed that allows the direct determination of *E. coli* in the Presumptive Test by the addition of MUG (4-methylumbelliferyl-B-D-glucuronide) to the media. Only *E. coli* can hydrolyze this molecule, yielding a fluorescent product. Thus, examination of the MPN tubes with a long-wave UV light will yield fluorescence in the presence of *E. coli*.

*Performing the MUG Test*

1. The MUG test will be set up as a demonstraion.
2. Examine the MUG tubes with the UV light, using a UV shield for protection.
3. Enter your results in the Evaluation of Results section and answer the questions.

# EVALUATION OF RESULTS
# (EXERCISE 26: WATER MICROBIOLOGY)

Purpose

Data

*Presumptive Test (MPN Determination)*

1. Record the number of positive tubes below.
2. Include results for your own group and for two other student groups.
3. Determine the MPN of coliforms present per 100 mL of water, according to the table in the exercise.

Number of Positive Lactose Broth Tubes

| Water Sample (source) | 3 tubes DSLB* (10 mL) | 3 tubes SSLB* (1.0 mL) | 3 Tubes SSLB* (0.1 mL) | MPN |
|---|---|---|---|---|
|  |  |  |  |  |
|  |  |  |  |  |
|  |  |  |  |  |

*DSLB = Double strength lactose broth;
SSLB = Single strength lactose broth

*Results of the Confirmed Test*

Record the results of the confirmed test for each of the above water samples.

| Water Sample Source | Positive | Negative |
|---|---|---|
|  |  |  |
|  |  |  |
|  |  |  |

*Results of the Completed Test*

Record the results of the completed test for water samples that were positive on the confirmed test.

| Water Sample (source) | Lactose Fermentation Results | Morphology | Evaluation |
|---|---|---|---|
|  |  |  |  |
|  |  |  |  |
|  |  |  |  |

## CONCLUSIONS, DISCUSSIONS, AND QUESTIONS

1.  Does a positive test mean that the water is absolutely unsafe to drink?

2.  What might cause a false-positive test?

3.  What advantage does the Millipore filter method have over the MPN method?

4.  List 3 characteristics of an organism that serves as a good sewage indicator.

5.  Why don't health departments directly test for pathogens, instead of using indicator organisms?

6.  List 2 bacterial diseases that are transmitted in polluted water.

    List 2 protozoan diseases that are transmitted.

# EXERCISE 27

## Spoilage of Meat

### A. SPOILAGE AT REFRIGERATION

### OBJECTIVES

At the conclusion of the exercise, you should...

1. understand why it is necessary to determine microbial counts on food.
2. be able to perform a standard plate count to determine the number of bacteria in a sample of meat.
3. identify some of the types of microbes that can grow at refrigerator temperatures.
4. understand why cooking meat is so important.

**Exercises required:**
Exercise 15 – Bacterial Plate Counts

### INTRODUCTION

Meats can become contaminated by microbes during and after the slaughtering process. Many contaminates come from the animal itself. Other contaminates come from the tools and utensils used in the preparation process. Freshly cut meats can have rapid bacterial growth. Immediate refrigeration after slaughter is necessary to slow the growth of the bacteria. Even though refrigeration slows the growth of bacteria, it does not stop it indefinitely. In time, certain bacteria will be able to grow at the cold temperatures. In this exercise, you will do a standard plate count on freshly bought and refrigerated meat and determine the number and types of bacteria commonly found in the meat. Within the last few years, a pathogenic strain of *E. coli* (O157:H7) has been isolated from partly cooked hamburger meat.

### MATERIALS

Safety: Biosafety Levels 1 and 2

Cultures:

Ground turkey (or beef) (10 grams per lab)

Media:

9 mL Sterile phosphate buffered water tubes
TSA plates
Bottle 90-mL phosphate buffered water


Supplies:

Bent glass spreading rods
Alcohol jar for flaming
Turntable per room
Bender (sterile)
5-, and 10-mL Sterile pipette canisters


## PROCEDURES
### Day 1 (Dilutions and Incubation)

*Technical Background*

Psychrophiles are bacterial species that will grow within a temperature range of -5°C to 20°C. The distinguishing characteristic of this group is that they will grow between 0°C and 5°C (refrigerator temperatures). Mesophiles will grow within the 20°C to 45°C temperature range. The distinguishing characteristic of the Mesophiles is their ability to grow at human body temperature (37°C) and the fact that they cannot grow above 45°C. Thermophile bacterial species will grow at 35°C and above. There are two groups of thermophiles: obligate thermophiles, organisms that will grow only at temperatures above 50°C, and facultative thermophiles that will grow at 37°C, with an optimum growth temperature of 45°C to 60°C. The psychrophiles are divided into obligate psychrophiles that seldom grow at temperatures above 22°C and facultative psychrophiles (psychrotrophs) that grow very well above 25°C. It is the group of low-temperature bacteria (psychrotrophs or low-temperature mesophiles) that causes the most meat spoilage during refrigeration.

Examples of psychrophiles include the Gram-negative genera *Aeromonas, Alcaligenes, Pseudomonas, Serratia,* and *Vibrio,* and Gram-positive genera *Bacillus, Clostridium,* and *Micrococcus.* Genera of psychrotrophs include Gram-negative genera *Acinetobacter, Citrobacter, Enterobacter, Escherichia,* and *Klebsiella.* Gram-positive genera include *Lactobacillus, Staphylococcus,* and *Streptococcus.* Most of these species are non-pathogenic, but there are some of the psychrotrophs that are significant pathogens. These include *Aeromonas hydrophilia, Clostridium botulinum, Listeria monocytogenes, Vibrio cholera, Yersinia enterocolitica,* and some strains of *E. coli.* In addition to the bacteria that can spoil meat, there are yeasts and fungi.

## Approximate Growth Temperatures

| Group | Range | Optimum | Examples of Habitat |
|---|---|---|---|
| Psychrophiles | -8–18°C | 15°C | Polar regions, deep ocean. * No pathogens of mammals or food. |
| Psychrotrophs Human pathogens: *Listeria, C. botulinum, Aeromena* | 0° to 30–40°C | 22°C | * Soil, plants, animals. Important in food spoilage (i.e., *Micrococcus cryophilus* and citrobacteria). |
| Mesophiles | 10–48°C | 25–40°C | Soil, plants, animals, water. Many species are human pathogens (i.e., *E. coli, Vibrio colerae, and Streptococcus pyogenes*). Some make enterotoxins when growing in food; ingestions may lead to food poisoning (i.e., *S.aureus, Bacillus cereus, and Clostridium perfringens*). |
| Thermophiles | 40–70°C | 50–60°C | Soil, hot springs, compost piles, endospore-forming. Some species may cause spoilage of food held at high temperatures (i.e., *B. stearothermophilus*). |
| Hyperthermophiles | 65–113°C | 80–105°C | Deep-sea hydro-thermal vents and volcanic hot springs. No pathogens or food-spoilage species. |

*Preparing the Inoculations*

1. The instructor will blend the 10 grams of meat in the blender with 90 mL of buffered water (1:10 dilution).
2. If necessary, the suspension will be filtered for easier pipetting.
3. Label the four tubes of water, 1-4.
4. Label the four plates 1:1,000 (1), 1:10,000 (2), 1:100,000 (3), and 1:1,000,000 (4).
5. Transfer 1 mL from the meat suspension to tube #1.
6. Perform a ten-fold serial dilution by transferring 1 mL from each tube to the next, through tube 4.
7. Plate 0.1 mL from tube #1 to plate #1.
8. Follow the diagram on the next page.
9. Place the plate on the turntable.
10. Dip the bent glass rod in alcohol and flame and cool.
11. Spread the 0.1 mL inoculum with the glass rod, turning the plate for even distribution.

12. Repeat the same procedure with all four dilutions.
13. Incubate one-half of the sets of plates at room temperature.
14. Invert and incubate the other half of the sets of plates in the refrigerator (4°C) for at least two weeks.

**Dilution Scheme**

1 mL meat transferred to tube #1, then a series of 1 mL
transferred from tube to tube (2 through 4):

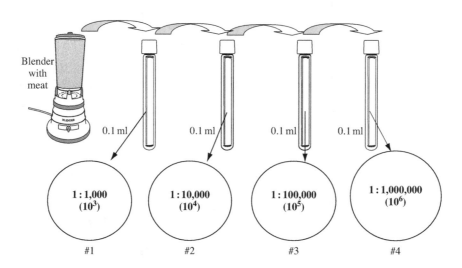

Blender with meat

0.1 ml     0.1 ml     0.1 ml     0.1 ml

1 : 1,000
$(10^3)$
#1

1 : 10,000
$(10^4)$
#2

1 : 100,000
$(10^5)$
#3

1 : 1,000,000
$(10^6)$
#4

# PROCEDURES
## Day 2 (After 2 Weeks of Incubation)

*Evaluating Your Results*

1. Count the colonies on each of the plates.
2. Select different colonies from each plate to Gram stain.
3. Record the results in the Evaluation of Results section.

## EVALUATION OF RESULTS
## (EXERCISE 27: SPOILAGE OF MEAT)

Purpose

Data

List and describe the organisms found growing on the plates.

| Colony # | Temperature | Morphology | Gram Stain |
|----------|-------------|------------|------------|
|          |             |            |            |
|          |             |            |            |
|          |             |            |            |
|          |             |            |            |
|          |             |            |            |

## CONCLUSIONS, DISCUSSIONS, AND QUESTIONS

1.  Discuss how cooking the meat before refrigerating might change the results.

2.  How might *E. coli* O157:H7 enter ground beef?

3.  Clearly differentiate between food poisoning by intoxication and food-borne infections.

4.  Autoclaving liquids will not kill some of the hyperthermophilic bacteria. Why is this not a problem for the medical profession?

5.  Complete the table below that gives food poisoning types and causative organisms.

| Name of organism | Food poisoning type | Symptoms in humans | Food commonly implicated |
| --- | --- | --- | --- |
| Botulism | | | |
| Staphylococcal | | | |
| *Clostridium perfringens* | | | |
| *Bacillus cereus* | | | |

# E X E R C I S E 28

## Microbiology of Wine-making

### OBJECTIVES

At the conclusion of the exercise, you should...

1. be able list different foods that microbes help produce.
2. make a sample of wine.
3. understand the process of fermentation used for producing a variety of foods.

### INTRODUCTION

Wine is essentially fermented fruit juice, wherein alcohol is produced using varieties of the yeast *Saccharomyces cerevisiae*. Wine is commercially prepared using a variety of grapes. Red wine is made from red grapes where the skin has been left on, and white wine is made from white grapes where the skin has been removed. The conditions necessary for wine-making include simple sugar, yeast, and anaerobic conditions. The yeast ferments the sugar to alcohol.

The word "wine" refers to any plant juice that is fermented by yeast. For example, dandelion flowers can be boiled in sugar water, and the resulting extract fermented to make "dandelion wine." A similar fermentation of freshly pressed grape juice results in wine. The yeasts convert the sugar in the juice to ethanol and $CO_2$, using glycolysis to make pyruvate, ATP, and NADH. The pyruvate is decarboxylated and reduced to ethanol, and the NADH reoxidized. Eventually, the ethanol concentration becomes high enough to stop yeast metabolism.

Vintners (wine-makers) go to great trouble to use just the right grapes, sugars, and yeasts to make their own particular wine, which usually has a pleasant taste. You will use generic materials to make wine, so your product will not taste like professionally made wine.

Microbial fermentations are used to produce food and beverages. Wine, beer, vinegar, buttermilk, cottage cheese, sauerkraut, pickles, and yogurt are all food products made from the fermentation of various microbes. In this exercise, you will make wine from fermented grape juice.

### MATERIALS

Safety: Biosafety Level 1

Cultures:

5-mL test tubes of various "starter" seed cultures of wine-making yeast

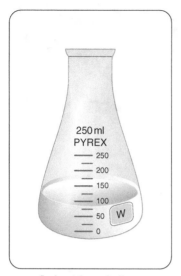

flask without balloon

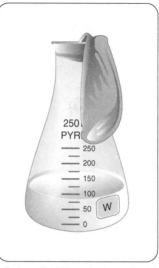

flask with balloon before inflated

flask with balloon inflated

## Media:

Grape juice – white or red (no preservatives)

## Supplies:

Air lock with cork
Balloons
250-mL Erlenmeyer flasks
pH paper

## PROCEDURES

### Day 1 (Inoculations)

*Technical Background*

Yeasts are unicellular fungi that usually divide by budding, i.e., the smaller daughter cell buds off from the larger mother cell. Most yeasts can be dried to a powder and stored in a cool place; when rehydrated in warm water, their activity returns rapidly. This is a great convenience for vintners and bakers. Yeasts are facultative organisms. If they have oxygen available, they prefer to use aerobic respiration of sugar to obtain energy. In the absence of oxygen, they switch to fermentation. Along with the addition of an air lock (with water), the liquid in this experiment is deep enough so that, in the absence of agitation, the lower levels become anaerobic. An **Air Lock** is a device (plastic or glass) that uses water as an insulator to protect the fermentation media from contamination and exposure to fresh air. At the same time, this device allows carbon dioxide produced by the yeast to escape the fermentation vessel. The air lock is also called a fermentation trap, or bubbler.

*Preparing the Inoculations*

1. Make a wet mount from the starter culture and examine it microscopically.
2. Measure 100 mL of grape juice into a 250-mL Erlenmeyer flask.

3. Add the rest of the 5 mL of starter yeast cultures.
4. Mix well by agitating the flask.
5. Measure the pH with pH paper. (Record the "before incubation" pH in the Evaluation of Results section.) Measure and record the height and circumference of the balloon.
6. Seal the flask with an air lock with cork (or a balloon).
7. If an air lock is used, add water to help promote anaerobic fermentation.
8. Incubate at room temperature (25°C) for at least 1 week.

## PROCEDURES
### Day 2 (Observations)

*Technical Background*

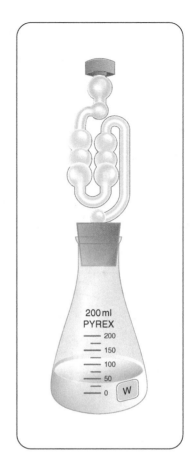

The overall equation for the alcoholic fermentation of sugars is $1\ C_6H_{12}O_6 \to 2\ CH_3CH_2OH + 2\ CO_2$. Carbon dioxide is quite soluble in water; it forms carbonic acid (HCOOH), which dissociates to H+ and bicarbonate. Thus, the pH of the solution drops. When no more $CO_2$ can dissolve, the gas comes out of solution and, in your experiment, collects in the balloon.

*Evaluating Results*

1. **If a balloon was used:** Measure the height of the balloon and its circumference. Record your results in the Evaluation of Results section.
2. **If an air lock was used:** Observe the water for bubbles.
3. Gently swirl liquid in flask.
4. Remove balloon or air lock.
5. Remove some liquid with plastic pipette.
6. Make a wet mount of the liquid and examine under 400X.
7. Make a drawing of the cells.
8. Find an area where the cells are evenly distributed in the field of vision and estimate their numbers.
9. Measure and record the pH of the liquid.

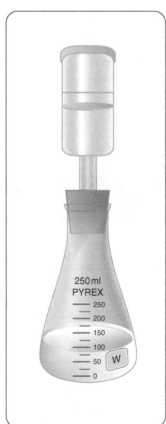

# EVALUATION OF RESULTS
# (EXERCISE 28: MICROBIOLOGY OF WINE-MAKING)

Purpose

Data

Length of Fermentation (days):

## Observations of Yeast Cells

| From starter culture tube (400X) | After growth in grape juice (400X) |

Number of cells
per field of view:_____                              _____

## Results from Fermentation of Grape Juice

| Character | At start | At end |
|---|---|---|
| pH | | |
| Circumference of balloon | | |
| Height of balloon | | |
| Aroma | | |

## CONCLUSIONS, DISCUSSIONS, AND QUESTIONS

1. List two places in nature where you might find yeast naturally occurring.

2. Why was the flask sealed with a balloon?

3. What is the purpose of the gas vent with cork?

4. Why is the pH important?

5. Once alcoholic fermentation ceases, the acetic acid bacteria can begin to grow, using the ethanol as an energy source. They need plenty of oxygen, as they are strict aerobes.

   a. What product results when this process is deliberately carried out?

   b. What can vintners do to prevent the growth of acetic acid bacteria in their wine?

6. Using your textbook, find the information to complete the table below:

| Fermentation end-product | Industrial or commercial use | Starting material | Microorganism (Domain) |
|---|---|---|---|
| Methane | Cooking gas | | Methanosarcina (Archaean) |
| Acetic acid | | Ethanol | |
| | Swiss Cheese | Lactic acid | |
| Lactic acid | | Milk | |
| Lactic acid | | Cabbage | *Lactobacillus* (Bacterium) |
| Ethanol | | Malted grains | |

## Microbiology of Milk

### A. STANDARD PLATE COUNT

### OBJECTIVES

At the conclusion of the exercise, you should...

1. determine the amount of bacteria in high- and low-quality milk, using the standard plate count method.
2. be able to perform the pour plate technique for making dilutions and plates for counting.
3. understand the pasteurization process.
4. be able to determine the difference between poor- and high-quality milk, using the reductase test for milk.
5. understand the fermentation process for making yogurt.
6. make and test yogurt.

### INTRODUCTION

The most reliable indication of the sanitary quality of milk is a bacterial count. The American Public Health Association has adopted the standard plate count method to determine the quality of milk. Even though pathogens may not be in high numbers, high counts may indicate a diseased cow, unsanitary handling conditions, or unfavorable storage. In this exercise, you will perform standard plate counts on two milk samples: a supposedly good sample (high-quality milk) and a poor sample (low-quality milk).

### MATERIALS

Safety: Biosafety Levels 1 and 2

Cultures:

15-mL sample of high-quality milk (pasteurized, refrigerated, whole milk)
15-mL sample of low-quality milk (raw, refrigerated milk)

Media:

99-mL sterile water bottle (for high-quality milk)
99-mL sterile water bottles (for low-quality milk)
TGEA melted pour tubes (in 50°C water bath)

Supplies:

4 Sterile Petri plates
1 mL Sterile pipettes
10-mL Sterile pipettes
50ºC Water bath

# PROCEDURES
## Day 1 (Inoculations)

*Technical Background*

**Pasteurization** is the name given to a method developed by Louis Pasteur to keep wines from sour-
ing. Today, the method is used for the treatment of all milk products. Pasteurization is accomplished
by heating to 60ºC for 20 minutes and holding at 10ºC after pasteurization. There is also a flash
method of pasteurization. In this process, the milk product is heated to 72ºC for 15 seconds, followed
by rapid cooling to 10ºC. These processes are used on other dairy products, such as cheese, ice cream,
yogurt, and vegetable and fruit juices to remove harmful bacteria and extend shelf life.

*Preparing Inoculations*

Work in groups, with one-half the lab assigned the high-quality milk, and the other half, the low-
quality milk.

### High-Quality Milk:

1.  Label the bottles and plates, as in the diagram.
2.  Make four pour plate count dilutions of the high-quality milk, as follows (see diagram).

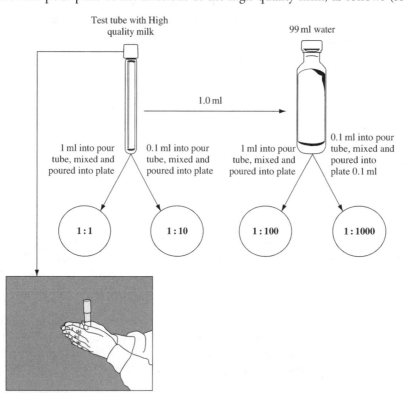

3. Shake the milk bottle well.
4. Transfer 1 mL from the original milk bottle into a melted and cooled TGEA pour tube.
5. Roll the tube between your hands and pour into the sterile empty Petri plate labeled 1:1.
6. Gently swirl the plate to distribute the agar evenly. Do not let the agar splash on or over the sides of the plate.
7. Transfer 0.1 ml from the milk bottle as above, into a pour tube, mix by rolling, and pour into the Petri plate labeled 1:10.
8. Pipette 1 mL of milk into a 99-mL water bottle (1:100); shake.
9. Pipette 1 mL from the 1:100 bottle into the agar as above, mix by rolling, and pour into the plate labeled 1:100.
10. Gently swirl the plate, as above.
11. Transfer 0.1 mL from the 1:100 bottle into the agar pour tube, mix, and pour into the plate labeled 1:1000. Swirl gently, as above.
12. Allow the agar to harden for at least 15 minutes.
13. Invert all plates and incubate at 35°C until the next lab period.

## Low-Quality Milk:

1. Label the bottles and plates as in the diagram in the high quality milk.
2. Using 99-mL water bottles and the standard plate count technique, as above (pour tubes) (see diagram).
3. Invert all the plates and incubate at 35°C until the next lab period.

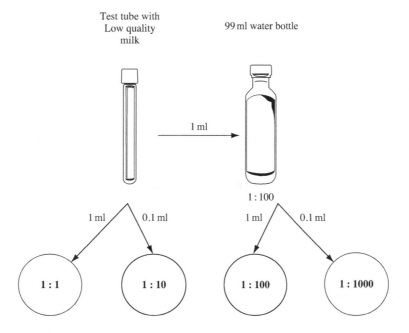

## PROCEDURES
Day 2 (Results)

*Technical Background*

Milk for sale to the public is graded according to the bacterial population obtained by the standard pour plate count method. **Certified raw milk** cannot exceed 10,000 bacteria/mL total, not to exceed 10 coliforms per mL. **Certified pasteurized milk** is not to exceed 500 bacteria/mL total, not to exceed 1 coliform per mL. **Grade A raw milk** is not to exceed 100,000 bacteria/mL, no more than 10 coliforms per mL, and **Grade A pasteurized** milk is not to exceed 20,000 bacteria/mL, not to exceed 10 coliforms per mL. Industrial-grade milk is not acceptable for drinking; it is used for the production of powdered milk, condensed milk, or evaporated milk.

*Evaluating Your Results*

1.  Count all the plates with colonies between 30 and 300.
2.  Record the results in the Evaluation of Results section.


## B. REDUCTASE TEST DEMONSTRATION

## MATERIALS

Safety: Biosafety Levels 1 and 2

Cultures:

Milk (different quality)

Media:

Different quality milk samples inoculated in 1-mL resazurin in 16 x 125 screw cap tubes, set up as a demonstration

Supplies:

37°C Water bath per room

# PROCEDURES (FOR INFORMATION PURPOSES)
## Day 1 (Inoculations)

*Technical Background*

Milk that contains large numbers of actively growing bacteria will become reduced due to the exhaustion of oxygen by the bacteria. Resazurin and methylene blue both lose color when they become reduced. This is the basis for the reductase test. In this test, resazurin is added to milk. The test tube is covered, mixed, and placed in a water bath at 35°C. The tube is examined at intervals up to 6 hours. The time it takes for the resazurin to change from pink to colorless is the basis of the reductase test. The shorter the reductase time, the lower the quality of milk. A reductase test time of 6 hours is good; 30 minutes is poor. In this exercise, the high- and low-quality milks will be tested for their resazurin reduction time.

*Checking Milk Resazurin Reduction Times*

1. Transfer 10 mL of the high-quality milk into a test tube with 1 mL of resazurin.
2. Repeat the same as above with the low-quality milk.
3. Cover the test tubes and gently mix.
4. Place the tubes in a water bath at 35°C.
5. Examine the tubes every 30 minutes, until the end of the lab period.
6. Record the results in the Evaluation of Results section.

# C. YOGURT PRODUCTION

# MATERIALS

Safety: Biosafety Level 1

Cultures:

1 cup of commercially prepared plain yogurt for starter culture

Supplies:

<u>per group:</u>
      250-mL Erlenmeyer flask
      Stirring rod
      Thermometer

<u>per room:</u>
      100-mL Cylinder
      Plain yogurt
      Powdered milk
      Whole milk
      Plastic wrap
      Weighing balances
      Weighing paper

Tongue depressers for spatulas
2 Boiling water bath setups
Gloves
42-45°C Incubator

## PROCEDURES
### Day 1 (Inoculations)

*Technical Background*

Yogurt is another example of a fermented food product. It is produced from the fermentation of milk products, using yeasts and lactic acid-producing bacteria. Usually, large-scale production of yogurt is done by adding pure cultures of *Streptococcus thermophilus* and *Lactobacillus bulgaricus* to pasteurized milk. In this exercise, you will make yogurt from milk products and use some commercial yogurt as the inoculum.

*Preparing the Inoculations*

1. Using a graduated cylinder, measure 100 mL of milk and transfer to a beaker.
2. Boil the milk, while stirring constantly.
3. Allow the milk to cool to about 45°C.
4. Add a teaspoon of the commercial plain yogurt provided (the inoculum).
5. Incubate at 42-45°C for 24 hours.
6. Transfer to a refrigerator until the next lab period.

## PROCEDURES
### Day 2

*Evaluating Your Results*

1. Evaluate your product's texture, color, aroma, and taste.
2. Make wet mounts.
3. Record your results in the Evaluation of Results section.

## EVALUATION OF RESULTS
## (EXERCISE 29: MICROBIOLOGY OF MILK)

Purpose

Data

**A. Standard plate count, and**

**B. Reductase test**

| Type of Milk | Plate Count | Dilution | Organisms/mL | Reductase test color/time observations |
|---|---|---|---|---|
| High-Quality | | | | |
| Low-Quality | | | | |

## C. YOGURT PRODUCTION

Record the following observations for your yogurt:

Color–

Aroma–

Texture–

Wet mounts:

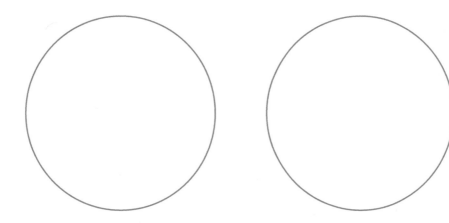

## CONCLUSIONS, DISCUSSIONS, AND QUESTIONS

1. After evaluating the standard plate count and reductase results on the two samples of milk, determine which milk is acceptable to drink.

2. Although human pathogens may not be present in high plate counts, what conditions might the high plate counts indicate about the improper handling of milk?

3. What advantage do you think reductase has over direct plate counts?

4. Does yogurt have any nutritional value? Why?

5. What type of fermentation is involved in yogurt production?

6. What is the most common source of bacteria in milk?

7. What infections can be transmitted to humans from milk?

8. Why is milk a more suitable vector of disease than water?

# EXERCISE 30

## Isolation of an Antibiotic Producer from Soil

## OBJECTIVES

At the conclusion of the exercise, you should...

1. isolate soil microbes that produce antibiotics.
2. learn how to make dilution plates using the spread plate technique.
3. learn about some of the different microorganisms found in soil.

**Exercises Required:**
Exercise 15 – Bacterial Plate Counts

## INTRODUCTION

Bacteria, fungi, and other microbes are found in soil. These microorganisms have an influence on the quality of life through their activities in soil by having roles in the carbon, sulfur, and nitrogen cycles in the environment. Also, many soil microorganisms produce antibiotics used in medicine. In this exercise, you will isolate bacteria from soil and test them for antibiotic production.

## MATERIALS

Safety: Biosafety Levels 1 and 2

Cultures:

*E. coli* - BSL1
*P. aeruginosa* - BSL2
*S. aureus* - BSL2
*S. epidermidis* - BSL1

Media:

Tubes of saline – 9 mL
Glycerol yeast extract agar plates (GYE)
TSA plates

Supplies:

Soil sample
Petri plate spreader
Alcohol jar for flaming
Sterile pipettes
Scales for weighing soil
Spatulas
Weighing paper

## PROCEDURES

Day 1

*Technical Background*

The group Actinomycetes encompasses a wide range of bacteria. They are Gram positive rods with with branching hyphae and specialized spore-bearing structures. Some of the Genera in this group include *Streptomyces* species, *Nocardia* species, and *Actinomyces* species.

*Nocardia asteroides* is found in the soil, and some strains have been pathogenic to man and animals. *Actinomyces israelii* is an anaerobe found in the human mouth, and can cause a disease called actinomycosis. The Streptomycetes are found in the soil, and there are many *Streptomyces* species that produce antibiotics. Some examples are *S. griseus*, which produces the antibiotic streptomycin, used to treat certain bacterial infections. *S. nivens* produces novobiocin, which is used to treat Gram-positive bacterial infections, and *S. orientalis* produces vancomycin, used to treat Staphylococci infections. These are only a few examples of the antibiotics that have been produced from the Streptomycetes found in soil.

*Preparing the Inoculations*

1. Label 6 x 9 mL saline test tubes 1-6.
2. Weigh out 1 gram of soil into tube #1.
3. Vortex or invert the tube to mix.
4. Make a tenfold serial dilution of the soil by transferring 1 mL from tube #1 into tube #2.
5. Mix each time after transferring.
6. Continue the serial dilution by transferring 1 mL from tube #2 through #6 (see diagram).
7. Label 3 GYE plates 1:10, 000, 1:100,000, and 1:1,000,000.
8. Transfer 1 mL from tube #4 (1:10,000) onto the plate labeled 1:10,000.
9. Dip the bent glass rod in the alcohol and flame.
10. Hold the rod a few seconds, until it is cool.
11. Spread the inoculum on the plate with the sterile bent glass rod, turning the plate several times for even distribution.
12. Repeat the procedure for tubes #5 (1:100,000) and #6 (1:1,000,000).
13. Invert the plates and incubate at 30°C for 1 week.

Each tube contains 9 ml of saline -1 ml is transferred from tube to

1 gram
soil

1   2   3   4   5   6

1 ml   1 ml   1 ml

**1 : 10,000**   **1 : 100,0000**   **1 : 1,000,000**

## PROCEDURES
Day 2

*Technical Background*

Streptomycete colony description: small colonies with pastel colors, including white, with a texture that is a hard mass that extends into the agar and sticks together tightly. Some of the mycelium grows into the agar, and the rest extends up from the agar, forming spores at the tips. The spores give the colony the pastel-to-white colors. They have a soil-like odor.

*Preparing the Streak Plates*

1. Examine the 3 GYE plates and find several colonies of Streptomycetes. (Note the colony description above, and compare your plate to the demonstration plates provided.)
2. Mark the colonies by drawing a circle around the colony on the bottom of the plate.
3. Each student will pick one of the colonies and streak it onto a TSA plate. Streak down the side of the plate (see diagram).

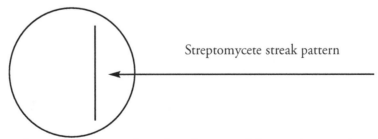

Streptomycete streak pattern

4. You will not be able to pick up the colony because of the way it grows into the agar, so use a cool loop and just scrape the loop across the top of the colony to pick up the spores.
5. Invert the plates and incubate at 30°C until the next lab period. This should give the Streptomycetes time to produce an antibiotic and diffuse into the agar.
6. The plates should be moved to the refrigerator if they are growing too fast.

# PROCEDURES
Day 3

*Preparing the Inoculations*

1. Examine the TSA for growth.
2. When growth is present, inoculate the TSA plate by cross streaking the four bacteria that are provided.
3. Streak up to the Streptomycete growth without touching it (see diagram).
4. Label the 4 streak lines as in the diagram.
5. Invert the plates and incubate at 35°C until the next lab period.

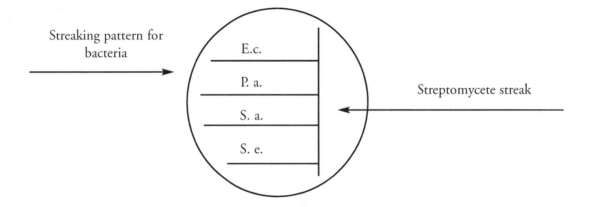

Streaking pattern for bacteria

E.c.

P. a.

S. a.

S. e.

Streptomycete streak

## PROCEDURES
Day 4

*Evaluating Your Results*

1.  Examine the TSA plates for evidence of antibiotic-producing *Streptomyces* species by looking for no growth (or halos) on the cross streaks (see diagram and demonstrations).

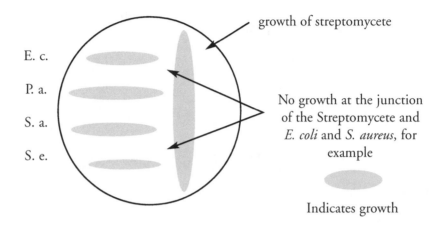

2.  Examine all the results in the class.
3.  Record the results in the Evaluation of Results section, and discuss the results in the Conclusion section.

# EVALUATION OF RESULTS
# (EXERCISE 30: ISOLATION OF AN ANTIBIOTIC PRODUCER FROM SOIL)

Purpose

Data

Describe the results in detail. Include the soil's origin (if known), and how many Streptomycete colonies were found on the original plates.

## CONCLUSIONS, DISCUSSIONS, AND QUESTIONS

1. Why was the soil diluted so much before plating?

2. If a Streptomycete inhibited growth of *Pseudomonas aeruginosa*, what group of bacteria would it probably inhibit?

# A P P E N D I X *A*

## Troubleshooting the Microscope

### *Some reminders for using the microscope:*

1. Always pick the microscope up with two hands: one under the base of the microscope and the other firmly grasping the arm.
2. Never turn the binocular eyepiece tube. It is set the best way for viewing.
3. Always clean the eyepiece and the objectives with lens paper before using the microscope, and be careful not to scratch the lenses.
4. The initial focusing of the slide is done with the coarse adjustment and the low power objective.
5. View the slide through the eyepiece with both eyes open.
6. Use the fine adjustment to obtain a sharper image.
7. Now switch to the high power objective and re-focus the slide.
8. Once in oil, only use the fine adjustment.
9. Never let the high-dry objective touch the oil.
10. When using live materials or wet mounts, it is best NOT to leave them exposed to the bright light for extended periods of time (it tends to dry out the slide and make it more difficult to view).
11. Remember, the object on the slide moves in the opposite direction when you are viewing it through the eyepiece (i.e., if you move the slide to the left, the object being viewed appears to move to the right).
12. When you have finished using the microscope, clean all of the lenses again.
13. Do not use any cleaner on the microscope objective. The objectives should be wiped with lens paper until all the oil is removed. If there is a problem, the instructor should try cleaning with alcohol and lens paper only. If it is still a problem, then bring the microscope to the prep room for repair.
14. Return the nosepiece to the low power objective position, and lower the stage.
15. Place the microscope in the cabinet facing in, arm toward door.

### *Some viewing and troubleshooting hints:*

1. If you cannot focus on the slide (or cannot find anything on the slide because it is very light), use a slide that has a heavy amount to focus on, then switch to the lighter slide. There are heavier slides available from other students or the instructor.
2. If you are having difficulty focusing the microscope, check the diopter adjustment (the two white dots should be lined up).
3. If you are still having difficulty focusing or finding anything on the slide, check to make sure the slide is not upside down.

# APPENDIX *B*

## Troubleshooting the Gram Stain

Some common problems, causes and their solutions, when performing the Gram stain:

| **Problem:** Can't find cells on the slide: | |
|---|---|
| **Causes:** | **Solutions:** |
| Objective lens is not focused on the dyed smear | Use 10X objective to locate smear; then rotate in the 100X lens |
| Cells were washed off during staining process | Draw a circle on the slide with a China marker or a Sharpie and prepare a fresh smear, air dry, heat fix, stain, and look for the stained area on slide within the drawn circle |

| **Problem:** Can find cells on the slide with the 40X objective, but not with the 100X one: | |
|---|---|
| **Causes:** | **Solutions:** |
| Not enough specimen on slide | Remake slide with more specimen (two loopfuls of broth) and/or use a QC slide (that worked) to get the correct field of view and then switch to the problem slide |
| Not enough play in the fine adjustment knob | Turn fine adjustment knob away from you about 6 turns; refocus with coarse, then fine, adjustment knob |
| Slide is upside down | Clean oil off slide by blotting with paper towel; invert slide and try again |
| Diopter is out of adjustment | Readjust ocular by matching the two white dots |
| 100X lens is dirty | Clean the lens gently with lens paper |

| **Problem:** Cells appear swollen and ruptured: | |
|---|---|
| **Causes:** | **Solutions:** |
| Normal cell appearance for the species | Confirm in wet mount |
| Too much heat applied during heat fixing | Prepare a fresh smear, air dry, and heat fix using less heat |
| Smear was not allowed to air dry completely, or the drying was hastened by heat | Prepare fresh smear and let dry without heat |
| Cells were subjected to hypotonic shock when suspended in tap water | Prepare fresh smear with saline |

| **Problem**: All the bacteria appear to be the same size and shape, but they stain pink and purple: | |
|---|---|
| **Causes:** | **Solutions:** |
| The culture is old | Prepare a new smear from a fresh culture |
| The smear is too thick | Prepare a new smear using fewer bacteria; use an inoculating needle instead of a loop |
| Bacteria may be acid fast | Prepare a new smear and perform acid fast stain |

# A P P E N D I X *C*

## Dilution Problems

Sometimes, we need to know the number of bacteria in a given sample. In many cases, there are so many cells in the original culture or sample that they cannot easily or accurately be counted. So, the sample must be diluted until a reasonable concentration of cells is reached and an accurate count is possible. The goal when diluting a sample is to reach a concentration of cells between 30 and 300 per milliliter.

### DILUTIONS:

Dilutions are performed by pipetting an exact amount of sample (usually 1 mL) into an exact amount of diluent (usually 99 mL). When 1 mL of sample is added to 99 mL of diluent, we have made a 1:100 dilution (read "one to one-hundred"). The first number, 1, refers to the volume of sample, and the second number, 100, refers to the volume of sample plus the volume of diluent (1 mL + 99 mL). Another way to think of this dilution is like this: One part of sample out of one hundred parts total.

> Example: You pipette 1 mL of sample into 9 mL of diluent. What is the final dilution?
> Answer: 1-mL sample + 9-mL diluent = 1:10 dilution.

### SERIAL DILUTIONS:

A serial dilution is a series of dilutions each made from the product of the previous dilution, where the bacteria are diluted in a stepwise manner. Thus, the concentration of bacterial cells proportionally decreases with each step of the dilution. Recall the 1:100 dilution from above. If we use 1 mL of that dilution and add it to another 99 mL of diluent, we will have performed dilutions in a series. The question arises, what dilution did we make when we added 1 mL of the 1:100 dilution to another 99 mL of diluent? A one to one-hundred dilution of a one to one-hundred dilution was made. In order to work with dilutions mathematically, we must think of them as fractions. Thus, a one to one-hundred dilution can be thought of as the fraction 1/100. When we make dilutions of dilutions, we calculate the new dilution by multiplying the original concentration as a fraction (in this case 1/100) by the amount we have diluted it as a fraction (in this case 1/100). So, the new dilution is 1/100 x 1/100 = 1/10,000 or 1:10,000.

> Example: You continue the serial dilution by pipetting 1 mL of the 1:10,000 dilution into another 99 mL of diluent. What is the new final dilution?

> Answer: a one to one-hundred dilution of a one to ten-thousand dilution was performed, which is: 1/10,000 x 1/100 = 1/1,000,000 or 1:1,000,000

## DETERMINING WHICH DILUTION YIELDS BETWEEN 30 AND 300 CELLS PER MILLILITER:

One mL and/or 0.1-mL aliquots from most dilutions in a series are plated out (grown on agar plates). Several plates should be made from each dilution, since this will give a better estimate of the actual concentration of cells present in each dilution. Once the cells have had time to grow, the plates are examined and it is determined which dilution has yielded between 30 and 300 colonies on a plate. Each colony is assumed to have grown from one cell, so the number of colonies is equal the number of cells. The colonies should be carefully counted and recounted. These colonies can also be referred to as colony forming units (CFU), and concentrations can be stated in CFU per mL. When 1-mL aliquots are plated, there is no calculation necessary to determine cells per mL (all colonies counted are from 1 mL). When 0.1-mL aliquots are plated, the count must be translated from cells per 0.1 mL to cells per 1 mL. Since 0.1 m is one-tenth of a mL, the number of cells counted must be multiplied by 10 to determine cells per mL. The term "plate count" (also "standard plate count" or "bacterial plate count") refers to this process of plating a known volume of sample and counting the colonies that grow from that volume to determine the concentration of cells.

## DETERMINING THE CONCENTRATION OF CELLS IN THE ORIGINAL SAMPLE:

Once we have determined the dilution necessary to yield between 30 and 300 cells per mL, we must figure out the concentration of cells in the original sample. Remember that a count from a diluted sample is not the same as a count from the original sample. If a serial dilution was performed until a dilution of 1:10,000 was reached, and it was plated out and it was determined that there were 120 cells per mL present in the dilution, then how many cells per mL were in the original sample? The following formula is used for this calculation. Remember that dilutions should be translated into fractions for calculations.

$$\textbf{Original concentration (cells/m)} = \frac{\textbf{Final concentration (cells/mL)}}{\textbf{Dilution (as a fraction)}}$$

So, original conc. in example above = $\dfrac{120}{(1/10,000)}$ = 1,200,000 = 1.2 x 10$^6$ cells/mL

(Note that the above equation can be used to solve for final concentration or dilution, as well as original concentration.)

## PRACTICE QUESTIONS:

1. You pipette 10 mL of sample into 990 mL of diluent. What is the final dilution?

2. You are performing a serial dilution. You have a 1:1,000 dilution, and you pipette 1 mL of that dilution into 999 mL of diluent. What is the new dilution?

3. You have a sample that you know contains approximately $4 \times 10^6$ cells/mL. If you make a 1:1000 dilution of this sample, what will be the concentration of cells in the dilution?

4. You perform a serial dilution on a sample until a final dilution of 1:1,000,000 is reached. By plating this dilution, you determine that the concentration of cells in the dilution is approximately 90 cells/mL. What is the concentration of cells in the original sample?

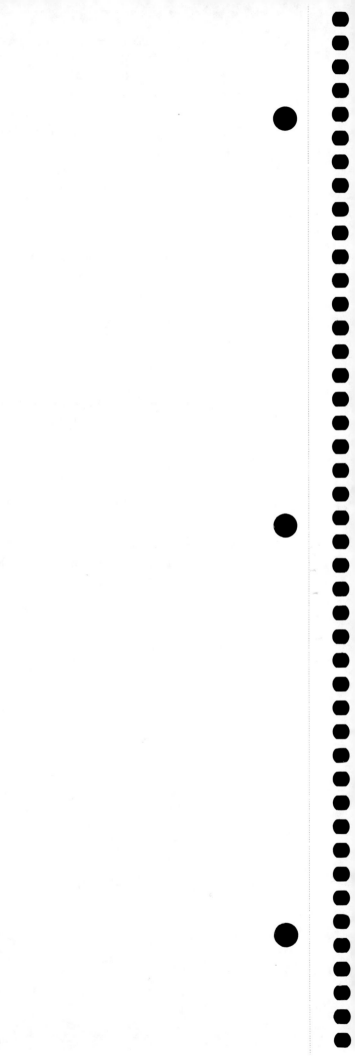

# A P P E N D I X *D*

## Flowcharts

Flowcharts are schematic drawings showing the sequence of operations to perform a task. In microbiology, they are commonly used for identifying unknown bacteria. For example:

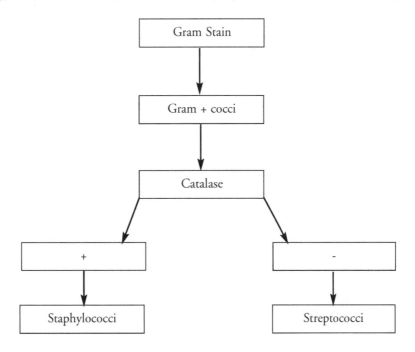

Also, just as a good table can collect and organize the data for results from the microbiology exercises, a flow chart will present a concise overview of the order in which things are to be done. Part of the preparation before coming to class is to look over the day's exercises and make flowcharts to help you to perform the laboratory steps. An example, using the throat culture exercise, is below:

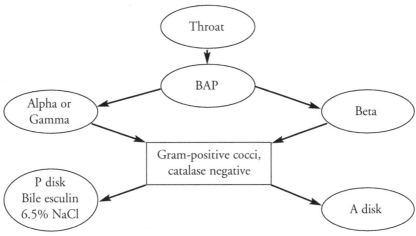

# APPENDIX *E*

## Unknown Identification Table:
## Gram-Positive Cocci - Catalase Positive

| Gram-Positive cocci | OXYGEN | TEMP °C | CATALASE | BCP LACTOSE | BCP MANNITOL | BCP RAFFINOSE | BCP SUCROSE | CITRATE | METHYL RED | NITRATE | COAGULASE | DNASE | HEMOLYSIS |
|---|---|---|---|---|---|---|---|---|---|---|---|---|---|
| *Staphylococcus aureus* | F | 35-37 | + | + | + | — | + | — | + | + | + | + | + |
| *Staphylococcus capitis* | F | 35-37 | + | — | — | — | — | — | — | + | — | W | — |
| *Staphylococcus cohnii* | F | 35-37 | + | — | — | — | + | — | — | — | — | — | — |
| *Staphylococcus epidermidis* | F | 35-37 | + | — | — | — | + | — | + | — | — | — | — |
| *Staphylococcus hominis* | F | 35-37 | + | + | — | — | + | — | + | + | — | — | — |
| *Staphylococcus saprophyticus* | F | 35-37 | + | + | + | + | + | — | — | — | — | — | — |
| *Staphylococcus sciuri* | F | 35-37 | + | + | + | — | + | + | — | + | — | — | — |
| *Staphylococcus simulans* | F | 35-37 | + | + | — | — | + | + | — | + | — | — | — |

**KEY:**

| O₂ Requirements: | Hemolysis: | Abbreviations: |
|---|---|---|
| Ae – aerobic | + = β ( Beta) | — = negative |
| F – facultative | — = none or γ (gamma) | + = positive |
| An – anaerobic | α = Alpha | W = weak positive reaction |
| Aan – aerotolerant anaerobe | w = weak alpha | V = variable (90% positive w/in 48 hours) |
|  |  | ND = not done |
|  |  | BCP = Brom cresol purple<br>+ = acid production |

# A P P E N D I X *F*

## Unknown Identification Table:
## Gram-Positive Cocci - Catalase Negative

| Gram-Positive Cocci | OXYGEN | TEMP °C | CATALASE | HEMOLYSIS | BILE ESCULIN | 6.5% NACL | Disk Susceptibility or Sensitivity Identification A | P | SXT |
|---|---|---|---|---|---|---|---|---|---|
| *Enterococcus faecalis* (Group D) | F | 35-37 | — | αlpha, w or gamma | + | + | — (R)** | — (R) | — (R)** |
| *Streptococcus agalactiae* (Group B) | F | 35-37 | — | βeta | — | V | — (R) | — (R)** | — (R) |
| *Streptococcus bovis* | F | 35-37 | — | αlpha, w or gamma | + | — | ND | — (R) | V** |
| *Streptococcus* equi (Group C) | F | 35-37 | — | βeta | — | — | — (R) | — (R)** | + (S) |
| *Streptococcus* mutans | F | 35-37 | — | αlpha, w or gamma | — | — | — (R)* ** | — (R) | + (S)** |
| *Streptococcus pneumoniae* | F | 35-37 | — | αlpha | — | — | V** | + (S) | ND |
| *Streptococcus pyogenes* (Group A) | F | 35-37 | — | βeta | — | — | + (S) | ND | — (R) |
| *Streptococcus salivarius* | F | 35-37 | — | αlpha, w or gamma | — | — | — (R)* ** | — (R) | — (R)** |

**KEY:**

| Hemolysis: | O₂ requirements: | ID disks: | Abbreviations: |
|---|---|---|---|
| α = Alpha | Ae = Aerobe | A = Bacitracin | R = resistant |
| β = Beta | F = facultative | P = Optochin | S = sensitive |
| γ = gamma | an = aerotolerant anaerobe | SXT = 1.25 mg trimethoprim plus 27.75 mg sulfamethoxazole | — = negative |
| w = weak alpha | Aan – aerotolerant anaerobe | + = Sensitive to reagent or drug in disc | + = positive |
| | | — = Resistant to reagent or drug in disc | W = weak positive reaction |
| | | * = some exceptions do occur | V = variable = + / — |
| | | ** = typically not done | ND = not done, or not significant |

# APPENDIX *G*

## Unknown Identification Table:
## Gram-Positive Rods

| Gram-Positive Rods | OXYGEN | TEMP °C | ACID FAST | ENDOSPORES | CATALASE | CASEIN HYDROLYSIS | NITRATE | METHYL RED | CITRATE 48 HOURS |
|---|---|---|---|---|---|---|---|---|---|
| *Bacillus cereus* | Ae | 30-32 | — | + | + | + | + | + | + |
| *Bacillus coagulans* | Ae | 30-32 | — | + | + | + | + | — | + |
| *Bacillus megaterium* | Ae | 30-32 | — | + | + | + | — | — | + |
| *Bacillus polymyxa* | Ae | 30-32 | — | + | + | V | + | V | — |
| *Bacillus pumilus* | Ae | 30-32 | — | + | + | + | — | + | + |
| *B. stearothermophilus* | Ae | 60-65 | — | + | + | ND | V | + | — |
| *Mycobacterium phlei* | Ae | 35-37 | + | — | + | V | + | — | — |
| *Mycobacterium smegmatis* | Ae | 35-37 | + | — | + | — | + | — | — |

**KEY:**

| O₂ Requirements: | Abbreviations: |
|---|---|
| Ae – aerobic | — = negative |
| F – facultative | + = positive |
| An – anaerobic | W = weak positive reaction |
| Aan – aerotolerant anaerobe | V = variable (90% positive w/in 48 hours) |
| | ND = not done |

# APPENDIX *H*

## Unknown Identification Table: Gram-Negative Rods

| Gram-Negative Rods | OXYGEN | TEMP °C | OXIDASE | ADONITOL | LACTOSE | MANNITOL | RAFFINOSE | SUCROSE | INDOLE | CITRATE | UREA | MOTILITY | METHYL RED | H2S |
|---|---|---|---|---|---|---|---|---|---|---|---|---|---|---|
| *Citrobacter farmeri* | F | 35-37 | — | — | — | + | + | + | + | V | V | + | + | — |
| *Citrobacter freundii* | F | 35-37 | — | — | + | + | V | V | — | + | V | V | + | + |
| *Edwardsiella tarda* | F | 35-37 | — | — | — | — | — | — | + | + | — | + | + | + |
| *Enterobacter aerogenes* | F | 35-37 | — | + | + | + | + | + | — | + | — | + | — | — |
| *Escherichia coli* | F | 35-37 | — | — | + | + | — | + | — | — | — | + | + | — |
| *Klebsiella oxytoca* | F | 35-37 | — | + | + | + | + | + | + | + | + | — | + | — |
| *Klebsiella ozaenae* | F | 35-37 | — | + | + | + | — | + | — | + | + | — | + | — |
| *Klebsiella pneumoniae* | F | 35-37 | — | + | + | + | + | + | — | + | + | — | — | — |
| *Morganella morganii* | F | 35-37 | — | — | — | — | — | — | + | — | + | + | + | — |
| *Pantoea agglomerans* | F | 35-37 | — | — | + | + | — | + | — | + | — | + | + | — |
| *Proteus vulgaris* | F | 35-37 | — | — | — | — | — | + | + | + | + | + | + | + |
| *Proteus mirabilis* | F | 35-37 | — | — | — | — | — | V | — | V | V | + | + | + |
| *Providencia alcalifaciens* | F | 35-37 | — | + | — | — | — | + | + | + | + | + | V | — |
| *Providencia rettgeri* | F | 35-37 | — | + | — | + | — | + | + | + | + | + | + | — |
| *Providencia stuartii* | F | 35-37 | — | — | — | — | — | — | + | + | V | + | + | — |
| *Pseudomonas aeruginosa* | Ae | 30-35 | + | — | — | — | — | — | — | + | + | + | — | — |
| *Pseudomonas aureofaciens* | Ae | 30-35 | + | — | — | — | — | — | — | + | — | + | — | — |
| *Serratia marcescens* | F | 35-37 | — | + | — | + | — | + | — | + | — | + | — | — |

**KEY:**

| O₂ Requirements: | Abbreviations: | |
|---|---|---|
| Ae – aerobic | + = positive | H$_2$S = hydrogen sulfide production |
| Aan – aerotolerant anaerobe | — = negative | MR = methyl red |
| An – anaerobic | V = variable (90% positive w/in 48 hours) | BCP = Brom cresol purple + = acid production |
| F – facultative | W = weak positive reaction | |

# A P P E N D I X *I*

## Writing Scientific Lab Reports

Communication of laboratory results and their significance is essential within the scientific community. Other scientists might just want to repeat your experiment in order to verify the results. The means of such communication is the **laboratory report,** and this is accomplished in a very specific scientific format.

The written lab report should answer the following questions:
1. What was the problem?      Your answer is the Introduction
2. How did you study the problem?      Your answer is the Materials and Methods
3. What did you find?      Your answer is the Results
4. What do these findings mean?      Your answer is the Discussion

Thus, the following sections should be included in the unknown lab reports. Each section should be labeled and should include the following:

## Title Page:
Includes title (Unknown Report Number ), Unknown number(s), Name, Class, Instructor, Section Number, Date.

## Introduction:
Explanation of problem (for example, state that you were given an unknown, what charts you used to identify the unknown, and why this was done).

## Materials and Methods:
The problem was studied using the culture methods learned in class. The physiological tests that were performed should be referenced here.

## Results:
Explanation of test theory (how and why it works and how a positive is detected).
Observation of results, not just positive or negative.
Gram reactions, colonial/cellular morphology, pigment production, etc., should be included.
Make a table to show the results.
Include a flow chart showing the positive and negative results.
This is where the answers are included, at the end of the flow chart.

## Discussion:
Discuss how the test results led to the identification.
Discuss any problems that might have occurred.
Include background information on the organism(s) that were identified (environmental/pathogenicity, etc.) Use other references to look up the identified organism, such as Bergey's manual.

## References:

Minimum required are from the laboratory manual and lecture textbook. If other resources are used then they should be included here, in the proper format. For example:

Benson, Harold J. (1994) *Microbiological Applications,* Wm. C. Brown.

## Worksheets:

These should be attached to the end of the report.

## Miscellaneous:

The report should have page numbers.

# APPENDIX J

## Selected Media and Reagent Descriptions

The following is a list of media that will be used in this laboratory course. Some media will be provided as demonstrations only. The purpose is to expose the student to a variety of media that is being used today in clinical and industrial microbiology laboratories. Media is used in solid (agar) and liquid (broth) forms:

**Agar [A]**: The dried substance extracted from marine algae or seaweed. It is used as a base for solid culture media and is usually not liquefied by most bacteria. It is ideal for culture media because it melts at 85° to 90°C and solidifies at 34° to 42°C. Also it is clear for ease of observing colonies and is not toxic to bacteria.

**Broth [B]**: A liquid culture medium that has essential nutrients for growing microorganisms and or specific chemicals added for specific testing of microorganisms.

## Types of media:

**General Purpose**: These are nonselective primary isolation media used for culturing a wide variety of microorganisms. Often these media are enriched with materials such as blood, hemoglobin, sera, or other growth factors.

**Selective Media**: Selective media is modified in a manner to suppress or prevent the growth of one group of microorganisms, while permitting the growth of another group from a mixed flora. The modification may be made by altering the pH or by the addition of a chemical that may be inhibitory to certain bacteria. Crystal violet will inhibit most Gram-positive types without affecting the Gram-negative group. Antibiotics may be used for selective inhibition. The degree of selectivity can be chosen also. Moderately selective media inhibits unwanted organisms to the point where they do not interfere with the growth of the wanted organism. Highly selective media strongly block the growth to completely eliminate undesirable cultures.

**Enrichment Media**: This type suppresses the growth of competitive normal flora, while enhancing the growth of the desired culture. Specimens known to be highly contaminated with normal flora often require exposure to enrichment media to be able to isolate the primary pathogen present.

**Specialized Isolation Media**: This group includes the media that is used to satisfy the nutritional needs of specific groups of organisms, thereby providing colonial morphology, differentiation, and identification. One example is Mannitol Salt Agar [MSA], which is used for the staphylococci group.

**Differential Media**: Bacteria display many variations in their nutritional requirements. One can take advantage of this by using carefully selected differential media for identification.

Some media can contain one or more fermentable carbohydrates and an indicator to detect acid or alkali productions. Some media contain substances rich in sulfur and an indicator to detect hydrogen sulfide formation; others contain a combination of both. Semi-solid media is used to demonstrate motility.

Some of the more common physiological properties determined on microorganisms and the media used are listed below:

**Carbohydrate Fermentation**: A carbohydrate is added to a basal medium (liquid or solid) to which an indicator has been added to detect changes in the pH that develop during growth. Production of gas is detected in liquid media by Durham tubes (small inverted vials that fill with liquid during sterilization). One example of a carbohydrate broth is one with a purple broth base - called Brom cresol purple (then the name of the carbohydrate); ex., BCP Glucose. Another example is to use Phenol Red broth base with carbohydrates. In both cases, a color change to yellow is a positive acid reaction. An example of a carbohydrate agar is EMB agar Base.

**Biochemical Substrate Utilization**: Another means of identifying organisms is by determining their ability to produce specific enzymes or metabolites, which, in turn, react with specific substrates. Culture media containing these substrates along with indicators are available. Some examples are Bile solubility, citrate utilization, and DNase.

**Media Descriptions**: Unless otherwise stated, most of the media are general purpose. They are listed alphabetically.

**Bile Esculin Agar [BE]**: Differential media primarily used for the identification of the Group D Streptococci. The Streptococci can tolerate the bile and some species have the ability to hydrolyze the esculin. Hydrolysis of esculin, which turns the agar dark brown, constitutes a positive reaction.

**Blood Agar Plate [BAP]**: Selective and differential type of medium. Trypticase Soy Agar [**TSA**] base is supplemented with sheep blood. The base medium is usually supplemented with 5% sheep blood. The TSA base is a nutritionally rich medium that supports good growth for a variety of microorganisms. The sheep blood is used for the isolation and cultivation of fastidious (fussy) microorganisms and for the identification of certain bacteria based on their hemolytic (lysis of red blood cells) reactions. The hemolytic reactions of the Streptococci groups are used as a preliminary identification before performing more tests. Blood Agar plates are used routinely for clinical specimens; therefore, it is important to know the hemolytic reactions of some of the more common bacteria, as well as the pathogenic ones.

Typical cultural responses after 18-24 hours at 35°C:
*Escherichia coli* - beta hemolytic
*Staphylococci aureus* - beta hemolytic and some strains are non-hemolytic
*Staphylococcus epidermidis* - non-hemolytic
*Streptococcus pyogenes* - beta hemolytic
*Streptococcus pneumoniae* - alpha hemolytic
*Streptococcus faecalis* - non- or alpha hemolytic

**Brain Heart Infusion [BHI]**: A highly nutritive medium used for cultivating a variety of fastidious microorganisms, including anaerobes.

**BCP Glucose [BCP]:** Media used for the detection of glucose sugar fermentation. Other examples of sugars used are **BCP Lactose, BCP Sucrose, BCP Mannitol**, plus many others.

**Chocolate Agar [Choc]:** Selective media that is used for the isolation and cultivation of fastidious microorganisms, especially *Neisseria* and *Haemophilus* species, from a variety of clinical specimens.

One of the ingredients is hemoglobin, which gives the agar an opaque brown color ="chocolate agar". Typical cultural responses after 24 - 48 hours at 35°C in $CO_2$:

> *Neisseria gonorrhoeae* - small, grayish-white to colorless, mucoid
> *Moraxella catarrhalis* - small, colorless

**Coagulase plasma:** Standardized rabbit plasma is used for detecting the coagulase enzyme produced by *Staphylococcus aureus*. If the organism produces the coagulase enzyme, then the plasma will clot. This indicates a positive coagulase test.

**Columbia CNA Agar [CNA]:** CNA agar supplemented with 5% sheep blood is a selective and differential medium used for the isolation and differentiation of Gram-positive microorganisms from clinical and non-clinical materials. The addition of the antimicrobial agents, colistin and nalidixic acid makes the medium selective for Gram-positive bacteria by suppressing the growth of Gram-negative bacteria, such as *Proteus*, *Klebsiella* and *Pseudomonas* species, while permitting growth of staphylococci, streptococci, and enterococci. The colistin disrupts the cell membranes of Gram-negative microorganisms, and the nalidixic acid blocks DNA replication in susceptible Gram-negative bacteria. Hemolytic reactions are determined from the sheep blood that is added.

**DNA agar [DNA]:** DNase test agar is a differential type of medium used for determining deoxyribonuclease activity of microorganisms, particularly *Staphylococcus aureus*.

**Eosin Methylene Blue [EMB]:** EMB agar is a selective and differential plating medium recommended for the detection of the Gram-negative reaction enteric bacteria. It utilizes the carbohydrate lactose. The eosin and methylene blue dyes in the agar inhibit Gram-positive bacteria to a limited degree. The dyes also help differentiate between lactose fermenters and lactose nonfermenters, due to the presence or absence of dye uptake in the bacterial colonies. Coliforms, as lactose fermenters, are visualized as blue-black colonies, whereas colonies of *Salmonella* and *Shigella*, as lactose nonfermenters, appear colorless, transparent or amber. *E. coli* colonies are very dark, almost black, and produce a green sheen when observed in reflected light. The green metallic sheen is produced from the very large amount of acid fermentation that produces a precipitation of methylene blue in the medium. This occurs with methyl red positive lactose-fermenting bacteria. *Enterobacter aerogenes* produce a dark center, often surrounded by a lighter color of pink, mucoid rim without the green sheen. This appearance occurs with lactose-fermenting bacteria that are methyl red negative. Another example is *Klebsiella* species). Non-lactose-fermenting bacteria appear as colorless colonies. Some Gram-positive bacteria, such as fecal streptococci, staphylococci, and yeasts, will grow and usually form pinpoint colonies. A number of nonpathogenic, lactose-non-fermenting Gram-negative bacteria will grow and must be distinguished from pathogenic strains by additional biochemical tests.

Typical cultural responses after 18-24 hours at 35°C:

> *Enterobacter aerogenes* - pink colonies, no sheen
> *Escherichia coli* - purple with black centers and green metallic sheen

*Klebsiella pneumoniae* - green metallic sheen, dark centers, usually mucoid
*Proteus mirabilis* - colorless colonies
*Pseudomonas* species - irregular, colorless colonies
*Salmonella* species - colorless
*Staphylococcus aureus* - inhibited (or slight growth)
*Streptococcus faecalis* - inhibited (partially)

**Glycerol yeast extract agar [GYE]**: Glycerol is a highly purified alcohol used as a carbon source for differentiating certain bacteria. The GYE agar plates are used for isolating and differentiating soil streptomycetes.

**Lactose broth**: A medium used for the detection of coliform bacteria in water, foods, and dairy products.

**mEndo MF broth**: A specialized medium used for the detection of coliforms in water using the Millipore filter method.

**M-staphylococcus broth**: A selective medium that contains sodium chloride. The sodium chloride will restrict the growth of many bacteria and, thus, select for staphylococci.

**MacConkey Agar [Mac]**: Mac is a differential plating medium recommended for use in the isolation and differentiation of lactose fermenting organisms from lactose-non-fermenting Gram-negative enteric bacteria. The differential action of Mac agar is based on the fermentation of lactose. Colonies of organisms capable of fermenting lactose produce a localized pH drop, which is followed by the absorption of the neutral red, and imparts a red color to the colony. Colonies of organisms that do not ferment lactose remain colorless and translucent.

Typical cultural responses after 18-24 hours at 35°C:
*Enterobacter aerogenes* - pink to red
*Escherichia coli* - pink to red with a precipitate around the colony sometimes
*Proteus vulgaris* - colorless
*Salmonella* species - colorless
*Shigella* species - colorless
*Staphylococcus aureus* - inhibited

**Mannitol Salt Agar [MSA]:** A selective medium used for the isolation of pathogenic staphylococci.

**Methyl Red broth [MR]**: A differential medium used for differentiating strains of coliform bacteria on the basis of acid production.

**Motility Stabs**: A semisolid gelatin agar used for demonstrating motility of microorganisms.

**Mueller Hinton [MH]**: Mueller Hinton is a general-purpose medium that is used for testing the susceptibility of microorganisms to antimicrobial agents. The agar plates are used in the disc diffusion technique.

**6.5% NaCl**: Salt broth that is modified to differentiate the enterococcal group D streptococci from the non-group-D streptococci. A positive Bile Esculin test and growth in 6.5% NaCl broth confirms the presence of enterococci.

**Nitrate broth**: Nitrate broth is used for the determination of nitrate reduction by bacteria. It is a valuable criterion for differentiating and identifying various types of bacteria. Certain bacteria reduce nitrates to nitrites only, while others are capable of further reduction to free nitrogen or even ammonia.

**Nutrient Agar and Broth [NA]**: A general-purpose media used for the cultivation of a variety of nonfastidious bacteria and for the enumeration of organisms in water, sewage, and other materials. Nutrient agar consists of peptone, beef extract, and agar. This relatively simple formulation provides the necessary nutrients for growing a large number of microorganisms that are not excessively fastidious.

**Phenylethanol Agar [PEA]**: PEA is a selective medium used for the isolation of staphylococci and streptococci from specimens containing Gram-negative reaction organisms, such as *Proteus* and *E. coli.* 5% sheep blood can be added for the simultaneous isolation of staphylococci and streptococci and the identification of their hemolytic reactions.

Typical cultural responses on PEA without blood after 18-24 hours at 35°C:
*Escherichia coli* - partially inhibited, very little growth
*Proteus vulgaris* - partially inhibited, very little growth
*Streptococcus* species - moderate amount of growth
*Staphylococcus* aureus - moderate amount of growth

**Peptone Iron agar [PIA]**: PIA agar is used for detecting the production of hydrogen sulfide ($H_2S$). It is used in differentiating members of the enteric family (Enterobacteriaceae).

**Potato Dextrose Agar [PDA]**: Is used for the cultivation and enumeration of yeasts and molds. The low pH (5.6) is favorable for the growth of fungi and yeast.

**Sabouraud Dextrose Agar [SAB]**: A selective type of medium used for the cultivation of dermatophytes and other pathogenic and nonpathogenic fungi from clinical and nonclinical specimens. The low pH (5.6) is favorable for the growth of fungi and inhibitory to contaminating bacteria in clinical specimens.

**SM 110 agar**: A selective medium containing a high concentration of NaCl for the isolation of staphylococci.

**Simmons Citrate Agar:** This agar is used for the differentiation and identification of bacteria (mostly in the family Enterobacteriaceae) on the basis of citrate utilization, citrate being the sole carbon source.

If the organism is capable of using citrate a the sole carbon source, then the agar will turn from the initial green color to a deep blue color. This indicates a positive reaction. A negative reaction is indicated by a green color (growth on slant, but no change in color).

**Skim Milk Agar [SM]**: Skim milk agar is used for the demonstration of coagulation and proteolysis of casein (casein hydrolysis). Casein hydrolysis is positive when the organism produces proteolytic enzymes that break down the casein and form a clear zone around the streaked area of growth.

**Thioglycollate Broth [Thio]**: Thioglycollate media is used for the cultivation of obligate anaerobes, microaerophiles, and facultative organisms. The incorporation of the reducing agent, sodium thioglycollate to the nutritional formulation of the media maintains a low Eh (redox potential) in the depths of the column of medium in tubes. Resazurin is an indicator of the oxygen tension of the medium. In the presence of oxygen, resazurin is pink; it is colorless in the absence of oxygen.

**Todd Hewitt [TH]:** A general purpose medium that is primarily used for the cultivation of the streptococci.

**Trypticase Soy Agar [TSA]:** A general-purpose media used for the isolation and cultivation of fastidious, as well as nonfastidious, microorganisms, including anaerobic, as well as aerobic, bacteria, although it is not the medium of choice for anaerobes. The combination of casein and soy peptones in the trypticase soy agar renders the medium highly nutritious by supplying organic nitrogen, particularly amino acids and longer-chained peptides. The sodium chloride maintains the osmotic equilibrium.

**Trypticase Soy Broth [TSB]:** The broth form of TSA that is described above.

**Tryptone broth:** Tryptone is a pancreatic digest of casein used as a nitrogen source in culture media for differentiating bacteria on the basis of their ability to produce indole.

**Urea slants:** This medium is used to detect urease production by bacteria.

## Specialized Reagents:

**Coagulase plasma:** Standardized rabbit plasma is used for detecting the coagulase enzyme produced by *Staphylococcus aureus*. If the organism produces the coagulase enzyme, then the plasma will clot. This indicates a positive coagulase test.

**Kovac's reagent:**  A special reagent used to detect indol. Once the reagent is floated on top of the culture (in tryptone broth), it begins to extract any indol present. The indol + reagent forms a bright red layer on the surface of the culture medium.

**Methyl red:** A pH indicator. It is red at pH values below 5.6. At pH values above 5.6, it is orange to yellow.

**Nitrite reagents A and B:** When added sequentially to a culture in nitrate broth, these reagents will react with any nitrite present and form a red color, the positive reaction. Wait at least 1/2 hour before recording the absence of color, which is a negative reaction.

# APPENDIX *K*

## Antibiotic Zone Diameter Interpretive Table

| Antibiotic | Zone Diameter Sizes (nearest whole mm) | | |
|---|---|---|---|
| | Resistant < mm | Intermediate mm range | Sensitive > mm |
| **AM** Ampicillin | Gram-neg. enterics: ″ 13<br>Staphylococci: ″ 28 | 14-16<br>– | ≥17<br>≥29 |
| **CB** Carbenicillin | Gram-neg. enterics: ″ 19 | 20-22 | ≥23 |
| **CF** Cephalothin | ″ 14 | 15-17 | ≥18 |
| **CIP** Ciprofloxacin | ″ 15 | 16-20 | ≥21 |
| **E** Erythromycin | ″ 13 | 14-22 | ≥23 |
| **Te** Tetracycline | ″ 14 | 15-18 | ≥19 |
| **NN** Tobramycin | ″ 12 | 13-14 | ≥15 |
| **VA** Vancomycin | Enterococci: ″ 14<br>Staphylococcus: – | 15-16<br>– | ≥17<br>≥15 |

**Antimicrobial Agent classification, disk code, and concentration (mcg = micrograms):**

**Penicillins:**
Ampicillin (AM) (10 mcg)
Carbenicillin (CB) (100 mcg)
Penicillin G (P10) (10 units)

**Cephalosporins:**
Cephalothin (CF) (30 mcg)

**Quinolones:**
Ciprofloxacin (CIP) (5 mcg)

**Aminoglycosides:**
Tobramycin (NN) (10 mcg)

**Macrolides:**
Erythromycin (E) (15 mcg)

**Tetracyclines:**
Tetracycline (Te) (30 mcg)

**Glycopeptides:**
Vancomycin (V) (30 mcg)

Note: The numbers on the discs refer to the concentrations of the antibiotic in mcg = μ (the discs are 6 mm in diameter).

References: BD BBL Sensi-Disc Antimicrobial Susceptibility Test Discs insert #88-4062-1, revised 2004/09

# A P P E N D I X *L*

## Spectrophotometer (Spec 20)

The spectrophotometer is used for determining bacterial growth counts by transmitting a beam of light through a bacterial suspension to a photoelectric cell. As the number of bacteria increases, the broth becomes more turbid, causing the light to scatter. Thus, less light reaches the photoelectric cell. The amount of light is measured as percent transmission or %T. %T is the amount of light passing through the suspension (or broth). Optical Density (OD) or absorbance is derived from the percent transmission. The OD is a logarithmic value and is used to plot bacterial growth curves.

Procedure for determining bacterial growth with the Spec 20:

### MATERIALS NEEDED:

> Blank (this is an uninoculated tube of the bacterial growth medium – typically, TSB)
> Spectrophotometer test tubes (13 x 100)
> Waste beaker

### PROCEDURE:

1. Pipette 4 mL of TSB into one of the 13 x 100 test tubes = blank.
2. Mark a line at the top of the tube with a Sharpie.
3. Follow the directions on the Spec 20:
   - Turn on – warm up 15 minutes
   - Set Zero
   - Set Wavelength (540)
   - Insert blank
   - Set Full Scale (100%)
   - Insert Unknown
   - Read %T or absorbance
4. Pipette 4 mL from the first growth tube to a clean 13 x 100 test tube.
5. Read the %T and the Optical Density from the meter.
6. Record results.
7. Dispose of the first culture in the waste beaker.
8. Pipette 4 mL from the second growth tube to be read, as above.
9. Continue this procedure until all tubes are read.
10. Results can be compared to the visual growth readings described in exercise 16 (0-3+).

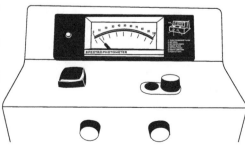

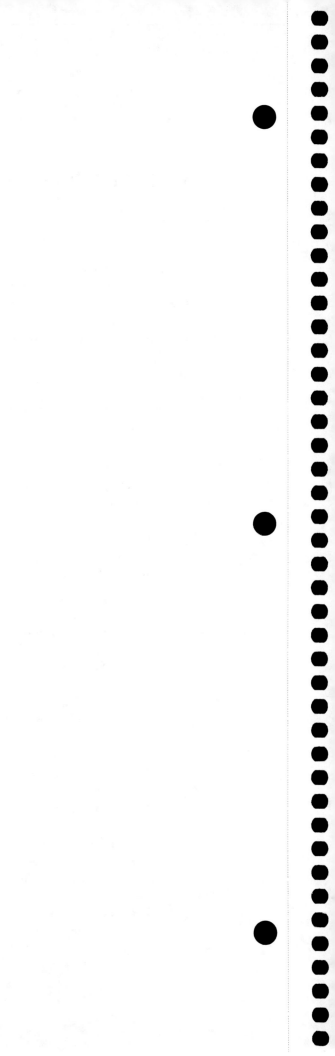

# A P P E N D I X *M*

## Pipetting

### I  SEROLOGICAL PIPETTING

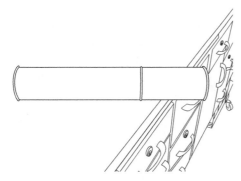

**Pipette Canister with sterile pipettes**

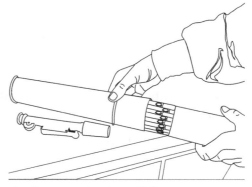

1.  **Remove canister lid.**

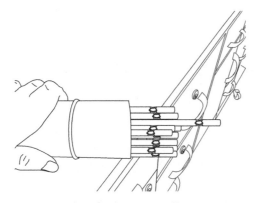

2.  **Use index finger to pull out one pipette.**

**DO NOT TOUCH OTHER PIPETTES!**

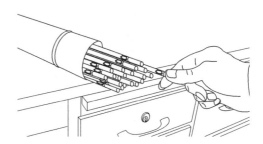

3.  **Aseptically remove one pipette.**

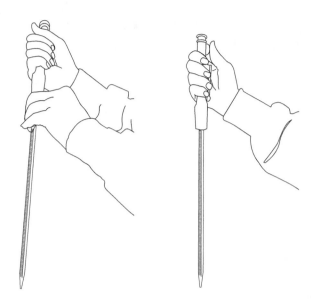

**4. Insert pipette into pipetting device.**

**5. Hold pipette in hand.**

## PROCEDURE:

- Remove top from canister.
- Aseptically remove pipette.
- Do not touch other pipettes.
- Hold pipette at top (lower portion is sterile).
- Insert pipette into pipetting device (push gently).
- Hold pipetting device in palm of hand, as shown.
- Use thumb and wheel for picking up specimen and transferring specimen.
- Carefully remove pipette by grabbing from top.
- Discard pipette into pipette jar (tip down).
- Replace top on canister.
- Never leave canister on end.

**How to read a seriological pipette**

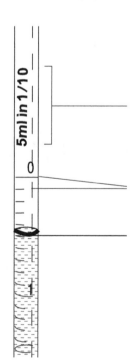

Indicates a 5-mL total volume pipette, divided into 1/10 or 0.1-mL increments.

Distance between these two lines = 0.1 mL.

Read amount at the bottom of the meniscus. This line is read as 0.5 mL.

## II MICRO PIPETTORS:

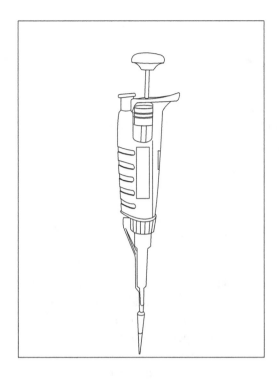

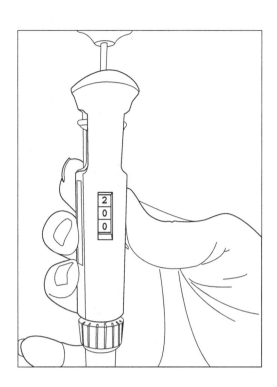

**P 20 Micro pipettor**                          **200 in the window = 20.0 μl**

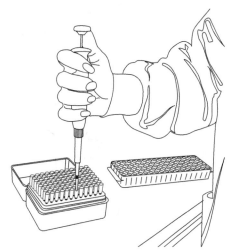

Press pipettor onto tip in pipette box.

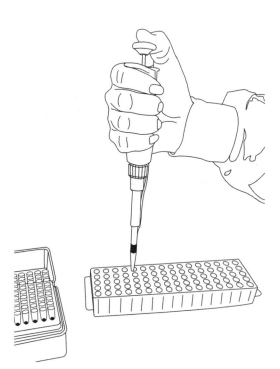

Press down on plunger to first stop. Then pick up sample by releasing plunger slowly. Then dispense liquid by pressing plunger slowly to first stop. Then press down all the way to blow out remaining sample from tip. When finished, eject tip into waste beaker.

# APPENDIX N

## Case Histories

The case histories presented here are meant to challenge students to develop a working knowledge of some of the different disciplines of medical microbiology that are taught in a microbiology course with a medical emphasis. These case histories have been taken from actual patient cases. By studying these case histories, students should better understand the importance of mastering the basic microbiology concepts and applying them in patient care.

**Directions:** Read the all of the following case histories and answer the questions using your text, any available reference books, and information from your lab exercises, as well as your notes from both the lab and lecture material.

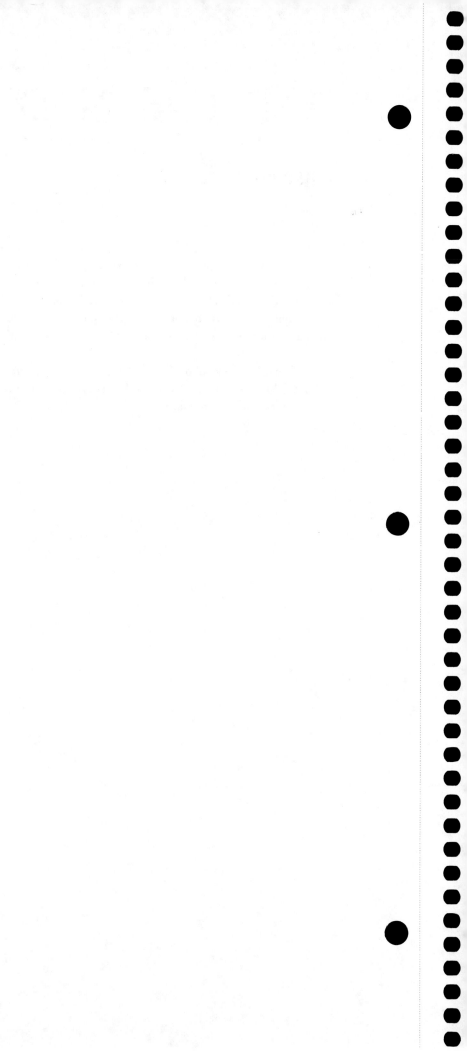

## CASE HISTORY #1

A 48-year-old man with a long history of alcoholism was admitted to the intensive care unit with gastrointestinal bleeding. He was intubated (tube inserted through the nose), and given intravenous fluids and a blood transfusion. He was treated with broad spectrum antibiotics. A culture was taken from his tracheal tube and initially showed *Staphylococcus aureus*. After further antibiotic therapy, a Gram stain of his tracheal aspirate showed polymorphonuclear leukocytes (White Blood Cells) and gram-negative rods. The culture of the tracheal aspirate this time yielded a heavy growth of an oxidase-positive, gram-negative, non-fermenting lactose rod that produced a greenish hue (color) on the culture plate and smelled like grapes. The patient was diagnosed with hospital acquired pneumonia.

1. What is the most likely bacterium (*Genus* & *species*) that caused this infection?

2. Is this organism frequently found as part of the normal flora of humans?
   Yes intestine, lungs or skin

3. Where can this organism be found in the hospital environment?
   Soil water from plants
   UTI infections → catheters.

4. What is the term given for this type of disease caused in the hospital?
   Called nosocomial → aquired in hospital

5. Is this organism normally sensitive or resistant to many of the commonly used antibiotics?
   resistant, didnt react to broad spectrum antibiotics

## CASE HISTORY #2

An 18-year-old male student went to the emergency room complaining of fever, chills, and pain while walking. He had many boils on his right leg, and he complained that they were painful. He had the boils for about one week and was hoping they would go away.

X- rays were taken of the right femur that showed some soft tissue edema (swelling) without any bone abnormalities. Bone scans revealed the possibility of acute osteomyelitis. Complete blood count (CBC) and chemistries were ordered. The WBC count was 15,000/μL and the differential showed 70% PMNs, 10% Bands, 20% Lymphs. All the chemistry tests were normal. Blood cultures and a wound aspirate were taken. The blood cultures were positive. A Gram stain of the wound aspirate and the positive blood culture revealed large amounts of Gram-positive cocci. The Gram-positive cocci occurred singly, in pairs, and in irregular grapelike clusters. They were catalase positive and were Beta hemolytic on blood agar. The colonies were yellow. The coagulase test was positive. Antibiotic sensitivity panels were done, and the organisms was resistant to penicillin and methicillin. A penicillinase-resistant type of cillin drug was given, and the patient was healed after four weeks of treatment.

1.   Based on the results given, what is the Genus and species of the pathogen?


2.   What widely known acronym is given to this species? (Hint: It has to do with an antibiotic.)


3.   List some of the reasons this patient could have gotten this infection?


4.   List other infections that are due to this bacteria.


5.   List three references used to answer these questions. At least one reference should be a textbook.

## CASE HISTORY #3

A 6-year-old patient with streptococcal sore throat takes penicillin for 2 days of a prescribed 10-day regimen. Since she begins to feel well, she is allowed to stop taking the penicillin. Three weeks later, she awakens with pain and swelling in the right ankle and fever. Upon evaluation by her physician, she is noted to have a heart murmur. She is admitted to the hospital with a presumptive diagnosis of rheumatic fever.

1. The patient had a sore throat that preceded her acute rheumatic fever. What pathogen would probably have been isolated from a throat culture before she was given penicillin? How can an "A" disc help in the identification of this organism? What is contained in the "A" disc?

2. Which type of hemolysis is seen when this organism is streaked on sheep blood agar?

3. How is pharyngitis with this bacterium related to the subsequent development of rheumatic fever?

4. What are some other types of disease caused by this organism?

## CASE HISTORY #4

**Background: A Case of Food Poisoning**

Twenty-two persons attended a brunch on Sunday, May 22, 1988 (11 a.m. to 3 p.m.) in Livonia, New York. All fresh food items were delivered to the cooking area the morning of May 22. The menu consisted of scrambled eggs, bacon, fruit salad, and pastries.

Scrambled eggs consisted of Grade A eggs from a Maryland farm. They were beaten, mixed with milk, salt, pepper, and diced onions. The eggs were scrambled in small batches and served from warming trays heated by alcohol burners that kept going out.

Bacon was purchased from a meat wholesaler. The bacon was fried on a large griddle and served from warming trays.

Fruit salad was made from fresh apples, bananas, and oranges. They were cut up and mixed with commercially canned peaches and pears.

Muffins and other pastries were purchased from a bakery and served with butter.

**Using the data in the table below, answer the following questions:**

1. Identify the etiologic agent of this outbreak of food poisoning.

2. Was it food-borne infection or intoxication?

3. How did the food get contaminated, and what item was contaminated?

4. Briefly explain how you arrived at your conclusions.

5. What other bacteria are associated with food poisoning?

Hints: 1. Make a summary table of the persons not ill.
2. Make a table of the onset of symptoms following eating.

**DATA**

| Case | Foods Eaten 1 | 2 | 3 | 4 | Beverages | Time of Meal | Symptoms | Onset of Day | Symptoms Hr |
|------|---|---|---|---|-----------|--------------|----------|--------------|-------------|
| 1  | x | x | x | x | C, J | 1100 | D, V, N, A | Sun | 2400 |
| 2  |   | x | x | x | M, J | 1200 |            |     |      |
| 3  | x | x | x | x | C    | 1200 | N, A       | Mon | 0200 |
| 4  | x | x | x | x | C    | 1200 | D, N, A    | Mon | 0600 |
| 5  |   |   |   | x | T    | 1200 |            |     |      |
| 6  |   | x |   | x | C, J | 1100 |            |     |      |
| 7  |   |   | x | x | C,J  | 1200 |            |     |      |
| 8  |   |   |   | x | C, J | 1300 |            |     |      |
| 9  | x | x | x | x | C    | 1300 | D, V, N, A | Mon | 1100 |
| 10 |   |   |   | x | T    | 1100 |            |     |      |
| 11 |   |   |   | x | M    | 1200 |            |     |      |
| 12 | x | x | x | x | M    | 1300 | D, N, A    | Mon | 1400 |
| 13 |   |   |   | x | C    | 1300 |            |     |      |
| 14 |   | x | x | x | C,J  | 1200 |            |     |      |
| 15 | x | x | x | x | T    | 1500 | D, N, V    | Mon | 2400 |
| 16 |   |   |   | x | M, J | 1500 |            |     |      |
| 17 |   | x | x | x | C    | 1100 |            |     |      |
| 18 |   |   |   | x | C    | 1100 | N          | Sun | 1300 |
| 19 | x | x |   |   | C, J | 1200 | D, V, N, A | Mon | 0100 |
| 20 | x | x | x | x | C    | 1100 | D, A       | Mon | 0800 |
| 21 | x | x | x | x | C    | 1200 | D, A       | Mon | 0200 |
| 22 | x | x | x | x | C    | 1200 | D, A       | Mon | 1200 |

Legend:  Foods eaten:  1-scrambled eggs,  2-bacon,  3-fruit salad,  4-pastry
       Beverages:  C- coffee,  T- tea,  M- milk,  J- orange juice
       Symptoms:  D= diarrhea,  N= nausea,  V= vomiting,  A= abdominal cramps

## CASE HISTORY #5

**Directions:** Read the following case history and answer the questions, using your text, any available reference books and information from your lab exercises, as well as your notes from both the lab and lecture material. (Note: questions are worth 1 point each)

A 30 year old sexually active woman was seen in the emergency room of a local hospital. She complained that she had burning while urinating and that it had been occurring for more than one week. A clean catch urinalysis with culture was ordered by the physician.

The UA results were as follows:

Glucose - negative
Bilirubin - negative
Ketones - negative
Specific gravity - 1.015
Blood - 1+
pH - 7.5
Protein - 1+
Urobilinogen - normal
Nitrite - positive
Leukocytes - 2+ (moderate)

Microscopically many red blood cells and white blood cells and many rod shaped bacteria were seen. The urine was sent to the microbiology lab and it was plated on EMB / BAP biplates with calibrated loops; 1 µl and 10 µl and incubated for 24 hours at 37°C.

Greater than 10,000 colonies (> $10^5$ CFU per mL) were seen on both biplates after 24 hours of incubation. The colonies had a green sheen on the EMB side and showed large grey colonies with slight hemolysis on the BAP side. The Gram stain result revealed Gram negative rods occurring singly and in pairs. The oxidase test was negative.

Antibiotic Sensitivity tests were performed with the standard antibiotics used for urinary tract infections caused by Gram negative bacteria.

1. List some of the bacteria that could be responsible for this UTI (urinary tract infection).

2. Which one of the bacteria is the most common cause of UTI 's?

3. How is this type of infection transmitted?

4. What antibiotics would be the most effective for treating this infection?

5. List three references used to answer these questions. At least one reference should be a textbook.
   These will vary, here are a few I use:

## CASE HISTORY # 6

**Directions:** Read the following case history and answer the questions, using your textbook, any available reference books and information from your lab exercises, as well as your notes from both the lab and lecture material and the Internet. (Questions = 0.5 point each, Total = 3 pts)

A 25 year old woman who was being treated for breast cancer went to her doctor complaining of a sore mouth and difficulty swallowing. The doctor examined her mouth and found white patches on the inside as well as on her tongue. The white patches looked like cottage cheese or milk curds.

The doctor was able to make his diagnosis based on the physical examination, but decided to take a culture and send it to the clinical laboratory for definitive identification.

The laboratory made a wet mount of the sample and observed clusters of budding yeast cells, and pseudohyphae. Pseudohyphae are a series of buds remaining attached to the parent cell and appearing as filamentous hypha. The laboratory was able to make a presumptive identification of the type of yeast so the doctor could start treatment. The doctor prescribed a topical imidazole treatment. The patient noticed improvement after two weeks.

1. What was the diagnosis that the doctor could make based on his observations?

2. What is the name of the yeast that was identified (Genus species)?

3. Where is this yeast normally found?

4. What is the most likely reason this cancer patient got the disease?

5. What is the drug of choice if the topical treatment did not work?

6. List the references used to answer these questions. At least one reference should be a textbook.

# APPENDIX O

## Selected References

Alcamo, I. E. *Laboratory Fundamentals of Microbiology*, 4th ed. Benjamin/Cummings Publishing Co., 1994.

Atlas, R.M. *Handbook of Microbiological Media,* 2nd ed. Boca Raton, FL: CRC Press, 1997. (The first edition may be consulted at the Prep Room window.)

Benson, H. J.; *Microbiological Applications, Laboratory Manual in General Microbiology*, 8th ed. McGraw-Hill, 2002.

*Difco Manual of Dehydrated Culture Media and Reagents*, 10th ed. Detroit, MI.: Difco Laboratories, 1984.

Eaton, A.D., et al. *Standard Methods for the Examination of Water and Wastewater*, 19th ed. Washington, D.C.: Am. Public Health Assoc., 1995.

Forbes, B.A., et al. *Bailey & Scott's Diagnostic Microbiology*, 11th ed. Elsevier Health, 2002.

Isenberg, H. D., ed. *Essential Procedures for Clinical Microbiology.* ASM Press, 1998.

Jay, J. M. *Modern Food Microbiology*, 4th ed. Chapman & Hall, 1992.

Leboffe, M. J. and Pierce, B. E. *A Photographic Atlas for the Microbiology Laboratory*, 2nd ed. Morton Publishing Co., 1996.

Sneath, P. H. A., et al. *Bergey's Manual of Systematic Bacteriology*, vol. 2. Baltimore, MD: Williams & Wilkins, 1986. (**Warning**: the taxonomy is out-of-date; the descriptions of species can be used.)

Tortora, G. J., et al. *Microbiology: An Introduction*, 6th ed. Menlo Park, Calif.: Benjamin/Cummings Publishing, 1997.

Bergey's Manual
Boone, D.R. and R. W. Castenholz, eds. *Bergey's Manual of Systematic Bacteriology*, 2nd ed., vol. 1, the *Archaea* and the deeply branching and phototrophic *Bacteria*. Springer, New York. (The most up-to-date taxonomy of the prokaryotes can be found in this volume.)

Holt, J.G., et al. *Bergey's Manual of Systematic Bacteriology*, vol. 1. Williams & Wilkins, 1984. (Covers many of the Gram-negative bacteria: spirochetes, non-motile curve bacteria, aerobic rods and cocci, facultative rods, sulfate-reducing bacteria, anaerobic cocci, rickettsias and chlamydias, the mycoplasmas. **Warning**: the taxonomy is out-of-date; the descriptions of species can be used.)

Holt, J. G., et al. *Bergey's Manual of Determinative Bacteriology*, 9th ed. Williams & Wilkins, 1994. (Covers many of the Gram-negative and Gram-positive bacteria. **Warning**: the taxonomy is out-of-date; the descriptions of species can be used.)

Sneath, P. H. A., et al. *Bergey's Manual of Systematic Bacteriology*, vol. 2. Williams & Wilkins, 1986. (Covers the Gram-positive bacteria: cocci, endospore-forming rods and cocci, regular nonsporing rods, irregular nonsporing rods, the mycobacteria, and the nocardioforms. **Warning**: the taxonomy is out-of-date; the descriptions of species can be used.)

# INDEX